Ball Lightning

Herbert Boerner

Ball Lightning

A Popular Guide to a Longstanding Mystery in Atmospheric Electricity

 Springer

Herbert Boerner
Mainz, Germany

ISBN 978-3-030-20785-4   ISBN 978-3-030-20783-0   (eBook)
https://doi.org/10.1007/978-3-030-20783-0

Cover illustration: Because ball lightening is so rare, photographs are few and far between and there are none that display the spectacular nature of this phenomenon well. To symbolize the idea of ball lightning, the cover photo shows a plasma globe, a device in which a barrier discharge is created in a noble gas atmosphere. The visual appearance may come close to that of an actual ball lightning.

This Springer imprint is published by the registered company Springer Nature Switzerland AG.
The registered company address is: Gewerbestrasse 11, 6330 Cham, Switzerland

Foreword

The physical origin, the stability, and the energetic maintenance of ball lightning have challenged lightning science for more than a century. The phenomenon is relatively rare, but eyewitness reports are common enough to inspire inquiry and motivate a scientific explanation. The author of this book, Herbert Boerner, a physicist who has worked on this subject for many years, makes a new effort to solve this riddle. His new book on this enigmatic subject with its bizarre behavior is rich in background. Detailed eyewitness accounts and photographic evidence on ball lightning are gathered and reviewed. The lightning physics relevant to this topic is extensively addressed, including appropriate emphasis on positive cloud-to-ground lightning, especially the exceptionally energetic variety that is called "mesoscale lightning" (and which is capable of ringing the Earth's Schumann resonances to intensities 20 dB greater than all the ordinary lightning combined). Laboratory experiments pertaining to ball lightning are explored, as well as reviews of all major texts on this subject. The greatest strength of this new work lies in its emphasis on the comparison between theory and observation, with a judicious use of Occam's razor. This pursuit is no small task given the author's estimate for 200 different existing theories on ball lightning. In the end, the author is turning to James Clerk Maxwell for an answer. The theoretical constructions that may be the basis for this phenomenon are relatively new, and so the book is timely in providing valuable new stimulus to both newcomers and professional scientists interested in solving this puzzle.

MIT
Cambridge, MA, USA
April 15, 2019

Earle Williams

Preface

It is interesting to look at the first few lines of the books on ball lightning. They usually start by referring to the first scientific study of ball lightning, which was published by the French scientist Arago in 1838 (Arago, 1838), showing that the subject has been around for a long time. Most of these books conclude with a hopeful sentence, stating that within a reasonable number of years, usually about 10, the author expects the problems concerning the physical nature of ball lightning to have been fully resolved. As it turns out, so far, these predictions have been much too optimistic. Ball lightning was and still is the most enigmatic form of atmospheric electricity. Arago's study is now 180 years old, but so far, there is neither an accepted theory nor a credible experimental study of these objects. Many scientists don't even believe it exists at all, due to its exotic properties.

So why should I write a new book on this apparently intractable subject? The answer is that I think it is no longer so intractable. In the 20 years that have gone by since the last book on ball lightning was published (Stenhoff, 1999), there has been a lot of progress in the understanding of lightning. Lightning location systems covering almost all regions of the globe are now in operation, giving us real-time information on the location and type of lightning occurring over land and also over the oceans. With the help of these location systems, a number of well-documented cases of ball lightning (or BL for short) have now been collected, giving us some hints on the circumstances under which these enigmatic objects appear and disappear and also allowing us to perform some thought experiments that could put theories about BL to the test.

Who is the intended audience of this book? There are two groups of readers, but people interested in BL with a basic knowledge of physics constitute the main readership. I have tried to keep the explanations as simple as possible (but not simpler), so everybody with a high school background in physics and mathematics should be able to follow the story rather easily. On the other hand, I have tried to be rigorous in applying well-proven scientific methods to eliminate unrealistic hypotheses, with which the field of BL research is unfortunately rife, so the readers with a scientific background but with a little knowledge of the field of BL research may also profit from reading this book. The information on BL is scattered over a number of books, some of which are difficult to get hold of, and also over a huge number of publications. Given the competitive atmosphere in science, most scientists will have been unable to get an overview of the topic of BL, and the book aims to provide a focused treatment with them in mind.

I begin the book by presenting several BL observations which I have collected from family members, friends, and acquaintances and which show that this interesting phenomenon is not so rare as many people believe. I then take a look at the photographic evidence and discuss some of the videos available.

One important aspect is the relation between what could be called mainstream science and BL. We will see the way natural science normally proceeds when analyzing new phenomena, and this will make it very clear why BL is such a hard subject for science. Then, we shall do a bit of philosophy, looking at an important tool called (rather fancifully) "Occam's razor."

The next chapter gives a short review of the earlier work on BL, but I will not attempt to duplicate the content of existing books. Almost all aspects of BL research were perfectly well covered in Mark Stenhoff's book *Ball Lightning: An Unsolved Problem in Atmospheric Physics* (Stenhoff, 1999). Yet his book was published in 1999, and new information on both BL and lightning in general has become available since then.

Lightning research has progressed a lot over the last 20 years, and since lightning is clearly the only way to produce BL, a whole chapter is devoted to the physics of electrical discharges, while another gives an overview of lightning physics.

Armed with this knowledge, the reader will then be ready for the discussion of some well-documented cases of BL observations which will allow us to draw rather strong conclusions concerning the nature of these objects.

Since Arago's work, there have always been scientists who are skeptical about the very existence of BL. However, a closer look at their arguments will show that their alternative explanations do not take sufficient note of the observational evidence and are often at odds with "Occam's razor."

The next chapter contains a critical review of the most popular theories of BL. As with the alternative explanations by skeptics, a check against the observations will allow us to eliminate almost all theoretical models.

I firmly believe that we have little chance of finding out what BL really is unless we can produce it in a laboratory in a repeatable way. Therefore, some experiments to generate BL, both accidental and intended, are presented. We will also see that we can define some parameters that may well be relevant for the creation of BL objects.

The last chapter wraps it all up. Finally, the appendix contains 26 original reports of ball lightning and bead lightning observations, that are relevant to the discussion in the book.

In my work on ball lightning, I have received help from many people, and without their support, this book would not have been possible. Donald Bäcker, Katja Näther, and Sven Näther shared their treasure trove of information on the Neuruppin event with me. Dr. Stephan Thern from BLIDS (Siemens) supplied me with information on lightning strikes, both for the Neuruppin case and also for my bead lightning observation. Regarding more recent investigations, Dr. Diendorfer from ALDIS kindly gave me the lightning strike data. I must also thank Dr. Diendorfer, Dr. Keul, Dr. Ute Ebert, and Dr. Earle Williams for the interesting and stimulating discussions. Over the last few years, I have exchanged literally hundreds of emails with Wilfried Heil and Peter Kocksholt who are collaborating with me on BL investigations; they have supplied me with information I could not possibly have obtained myself. I also have to thank the people that allowed me to use their photos in this book, see the figure captions for details. And last but not least, I would like to acknowledge the help of many of my friends and acquaintances who have shared their observations of ball lightning with me.

Mainz, Germany   Herbert Boerner

Contents

1

Introduction

My interest in ball lightning dates back 40 years. When I was a physics student, I traveled around Australia as a backpacker, staying in youth hostels. The place I liked most was called "Lost World Youth Hostel", close to Lamington National park in southern Queensland. It was there that I bought a book called "Green Mountains" by Bernard O'Reilly (1962), describing the search for the survivors of a plane crash in 1937, near the location of the youth hostel, and the pioneer's life in the rain forest on the border ranges between Queensland and New South Wales.[1] Bernard was obviously a brilliant observer of nature, and he gave accounts of many interesting events, including the total solar eclipse of September 1922. In these mountains, in November of the same year, he witnessed what he called an "electric thunderstorm", which produced a considerable number of BL objects: luminous balls, glowing in a deep red, like the embers of a log fire. When I read this account, I immediately got hooked on these mysterious objects, which I had previously regarded as extremely rare oddities. However, this report quite convincingly showed that, under the right conditions, ball lightning events can be produced in quite large numbers. This suggests that it may be possible to generate such objects in the laboratory, giving a real chance to study their properties in detail. Back at university, I talked to my professor, who told me immediately that it would be much more worthwhile to concentrate on my PhD thesis than such exotic stuff. Well, I followed his advice (or request, I should say) but I never forgot about Bernard O'Reilly's account.

Many years later, when I managed to get some free time (our children had grown up by then), I started to investigate other ball lightning reports and

---

[1] The O'Reilly family still runs a holiday resort in Lamington National Park.

© Springer Nature Switzerland AG 2019
H. Boerner, *Ball Lightning*, https://doi.org/10.1007/978-3-030-20783-0_1

I obtained literature on the subject. Around the year 2000 I stumbled across a particularly interesting case which had happened at Neuruppin in eastern Germany in 1994, where no fewer than 11 ball lightning objects were seen in a brief winter thunderstorm. This case turned out to be extremely well documented, first by the staff of the local meteorological station, and then by a couple who interviewed the witnesses about 1 year later. When I was able to obtain data about the thunderstorm from a lightning detection network, a completely new twist to the story emerged and—as you will see later—it helped to clarify several questions concerning these enigmatic objects. A few years ago, I retired, and since then I have been able to devote more time to my hobby. The book is organized around accounts of observations which are currently our only sources of knowledge on ball lightning.

2

Ball Lightning: Observers' Tales

Some Reports to Whet Your Appetite

There are more people who have seen ball lightning then one would think. It's not so rare at all. During the last two years, I have been talking about my favorite hobby to most of my friends and acquaintances, and I was surprised to find out just how many have seen such objects. The count now stands at 12, and I hope to be able to add more soon. So let's have a look at some examples. These accounts will explain what ball lightning looks like much better than I could. We start with a case associated with the most enigmatic circumstances, because it took place in a modern all-metal plane.

"It was certainly before 1992, but I cannot remember the exact year. I was working as a flight attendant for BA, flying on a BAC One-Eleven. We had reached our cruising altitude and we three (me and my two colleagues) were standing in a circle in the small kitchen in the front of the airplane, me with my back towards the cockpit, when suddenly a bright golden sphere of about 40–50 cm diameter appeared amongst the three of us. I think it appeared just there, but it might also have come from the cockpit. It shot down the aisle towards the rear of the plane. We three saw it for sure, but I cannot remember if any of the passengers saw it as well. There cannot have been a thunderstorm outside, because then we would have been sitting with the seat belts fastened. I was sure that this was ball lightning, because when I was a little girl, I had already seen ball lightning. I was outside, fetching something for my father on a thundery day, with sheet lightning, when about 50 m away to the right a golden ball of maybe 1.5 m diameter appeared, moving fast in my direction."

Two acquaintances told me the following story:

© Springer Nature Switzerland AG 2019
H. Boerner, *Ball Lightning*, https://doi.org/10.1007/978-3-030-20783-0_2

"It was in Patagonia in 2008. Our car was standing on a dirt road on the Valdez peninsula. There was a thunderstorm going on, with lightning in the clouds but no lightning to the ground. There was no rain falling near the car. Suddenly a yellow ball of about 20 cm diameter fell vertically from the sky, hit the ground about 2 m away from the car, passed underneath it and hit a bush 100 m away, which burst into flames". My acquaintances were absolutely sure the bush was on fire after the ball had vanished.

Recently, I asked a good friend of mine about ball lightning and I was quite surprised to hear that she also had a story to tell:

"We—me and a friend of mine—were standing at the window watching the clouds and the frequent lightning of an unusually strong thunderstorm. The house was on a hill above the town, on a road leading downhill towards the center. Suddenly a yellow ball appeared on the road, moving fast downhill, faster than a car. Its intensity was blinding, it had a diameter of at least half a meter and its outer surface was not sharp, it looked like frayed or sparkling. It followed the road like a bicycle. I remember it clearly because we quarreled about where precisely it had vanished on its race downhill. I was about 12 or 13 years old and it was in a town called Gunzenhausen in southern Germany."

Last but not least, my mother-in-law had also seen one of these objects:

"I was living in a small village north of Berlin. On the afternoon of a hot summer's day in 1934 or 1935 I was swimming in the lake close to the village, when a thunderstorm came up. I rushed home on my bicycle, but the storm was quicker and rain and thunder caught up with me. On the right side of the rough village road was a fenced meadow. Suddenly, after a loud thunderclap, a bright red or orange/yellow ball approached me, rolling along the top wire of the fence. This round lightning was much smaller than a football, and its surface was not smooth, but it had many spikes which emitted sparks. I hasted home to tell my parents what I had seen. My father said it must have been ball lightning."

All these reports are typical of ball lightning observations. Usually these "balls of fire" are seen in a thunderstorm, but there are exceptions. In the accounts above, the ball lightning objects appear in the open, but they are often seen inside houses. This leads to some of the most puzzling reports, like those I collected from the Neuruppin case (which I will explain in detail below):

"I was sitting in my favorite corner of the living room, with my daughter playing around near the armchair. Suddenly, there was a bright light between me and the child. At first, I thought that the child had been playing with the camera, which was lying on the armchair, but she looked at me with a frightened expression. From the hallway, my husband was also looking at the light with a stunned expression. It lasted only 1–2 seconds. It was directly over the floor, oval-shaped, like a large egg, and about 1 m long. In the middle it was blue and around the outside it was very bright. It didn't move…. Nothing was damaged. Outside it started to rumble and then there was a very loud bang." The son saw a similar object in a neighboring room, but it was of a yellowish color.

"I was sitting in the corner of the room, reading a newspaper. Suddenly something swished past me, coming through the closed window. The net curtains were closed, but the main curtains were open. The object had the size of a small glass lampshade and looked like a yellowish light bulb. From the window it moved to the opposite side of the room along the cupboards to my left, then it made a U-turn and came back straight towards me, but luckily it didn't touch me. It passed me at a close distance and went back out through the window. […] Its movement was very swift. It hissed, and the room was more brightly lit than by sunshine. […] I did not hear any thunder. The curtain was undamaged, as was the TV set."

The ball lightning in the first report appeared inside the house basically "out of thin air". It didn't move and was gone quickly without leaving a trace. In the collection of sightings from Neuruppin, there is another case like this, and reports from other sources confirm this observation. In the second report, the object entered through a closed window and a curtain, made a round trip of the room, and went out through the window again. Such a behavior is fairly often reported. When someone who has not seen ball lightning hears such reports for the first time, the reaction is often "this simply cannot be true, such a thing cannot exist". The reaction of most physicists is similar, and even more vehement, but they also have some arguments to support their point of view, as we will see later. On the other hand, witnesses are often perfectly trustworthy (as is the case for those quoted above), so we can exclude the idea that they were simply inventing a story to puzzle the scientific community. But what did they see then? Was it really something physical or was it a hallucination? Or something preternatural, as some people claim?

Did They Really See Ball Lightning or Was It Something Else?

Several other natural phenomena have characteristics that may appear similar to the objects in the accounts given above. In the nineteenth century or earlier, meteorites were often confused with ball lightning. When meteorites enter the atmosphere, this sometimes produces brilliant fireballs that travel for hundreds of kilometers before exploding. We now understand the origin of these bolides, but this was not known before the beginning of the nineteenth century, and initially it was not widely accepted. In fact, Thomas Jefferson is supposed to have said: "I would rather believe a Yankee professor would be lying than stones would fall from the sky".

Bolides vanished from the ball lightning reports at the beginning of the twentieth century when their origin was clear, but then another and more serious source of confusion appeared: electrical power lines. When lightning strikes power lines, electrical arcs may appear at various points along the line, for example, across insulators or at over-voltage arresters. These arcs can be very bright and may sometimes look more or less spherical, especially if viewed from a distance. There is one photo of such an arc producing a luminous mass of air which was blown away by the wind and could be seen for a few seconds (Stenhoff, 1999).[1] Some storm chasers call these arcs "flashballs". When they are not insulated, the bare wires of power lines may also touch and produce arcing between the different phases. These arcs can travel along the lines for some distance. Quite spectacular videos of such events can be found on the net, often erroneously labeled as "ball lightning observations".

Normal lightning can also adopt strange shapes which may mimic a more or less spherical source of light, but this is quite rare, and it only applies to objects that appear high in the sky.

Since WWII, aliens would appear to be visiting us in their spacecraft in amazing numbers. Of course, the alleged observation of UFOs has been linked to ball lightning and vice versa. Yet there is a fundamental difference between UFO and ball lightning observations: whereas the ball lightning objects have a quite consistent description, the UFOs seem to appear in a bewildering variety of forms.

When we compare the reports above with these possible sources of misidentification, it is clear that none of the cases can be attributed to them. Neither bolides nor high-voltage arcing can be held responsible for the observations. But is there something else that we might have missed? People interested in ball lightning have compiled a list of additional phenomena that could be the culprit (Stenhoff, 1999), but basically none of them could possibly be linked to our reports except one: corona discharge or St. Elmo's fire. In order to explain what this is, I can supply an observation of my own. Unfortunately, so far, I have never seen ball lightning, but I was once lucky

---

[1] A more recent example of multiple "flashballs" can be found here: http://foudre.chasseurs-orages.com/viewtopic.php?t=6313&highlight=

enough to observe St. Elmo's fire. In early October 1980, I visited Santorini island in the Aegean Sea. For almost a week the weather was perfect, but in the morning when I was about to fly back to Athens, a cold front approached the island from the north. I was waiting for the bus to the airport at a small road in the countryside when I saw a dark and ominous shelf cloud approaching at great speed. Lightning fell from it onto the countryside below. I hoped that the bus would arrive before the thunderstorm, but the clouds came faster. Just before the shelf cloud was overhead, I heard a strange humming noise. Looking up, I saw yellow bundles of flame on the insulators of a power line. They were dirty yellow in color, writhing around and humming and hissing. I knew what it was and that it signaled the existence of a very high electric field produced by the thunderstorm clouds, threatening me with a close lightning strike. I looked for shelter, but the house nearby was locked, and its entrance porch provided only very little protection. Seconds later, torrential rain came down, and the flames slowly went out. I was soaked through by the rain and I missed the bus, but since the flight was several hours late due to the bad weather, I did not miss the plane.

Electrical discharges like this are often seen in places where the electric field in the atmosphere is very high. These discharges are observed at sea, in particular, where a ship is the largest object above the waves for many miles around. Sailors, being eminently superstitious, regarded them as a sign from heaven indicating that the worst of the thunderstorm was over, but this would appear to be a misconception, because they clearly indicate that the electric field of the thunderstorm is still very high. Airplanes also sometimes display this phenomenon.[2]

This type of discharge is clearly related to the electric field created by the thunderstorm. The color is usually blue,[3] but in my case, the insulators may have been contaminated with salt from the sea nearby, giving the yellowish appearance due to the sodium vapor present in the discharge.

In several books or publications on ball lightning, it is claimed that this discharge can be confused with BL because it takes the form of a glowing ball. I would tend to disagree with this statement, since I have not been able to find a single description of such an observation in which corona discharge actually takes this form. Corona discharge is always described as consisting of a bundle of filaments, called streamers, emanating from sharp points of objects, usually

---

[2] Many images on the web alleged to show St. Elmo's fire on cockpit windows are actually discharges on the surface of the window and not a corona discharge.

[3] The blue color is due to excited nitrogen molecules in the air.

conductors. A discharge like this can only take the form of a sphere if it is starting from a round conductor like a door knob, or something similar.

At this point, readers will probably ask themselves whether there is any independent evidence of the existence of BL, like photos or videos, which can resolve the problem. And if so, why doesn't he show it? Indeed, there are photos and videos, but few are clearly authentic, and they do not show all the known features of BL.

Summary

- I am surprised how many of my friends and acquaintances have seen BL. The reports are from people I can trust, so there are no "fake reports".
- Confusion with other natural phenomena like St. Elmo's fire can be excluded in all these cases, since St. Elmo's fire clearly differs from what is reported as BL.
- Up to the nineteenth century, bolides streaking across the sky were confused with BL.
- In the twentieth century, arcing at high voltage lines became a new source of confusion.
- Since the end of WWII, UFO sightings abound and some of them may be due to BL objects.

3

The Search for Photographic Evidence

The search for photos or videos of ball lightning is an endless story of confusion and obfuscation, starting basically with the widespread availability of cameras and photographic plates in the early twentieth century and continuing up until today with powerful video recorders at everybody's fingertips. When photographers tried to catch photos of normal lightning, they discovered a technique which is still used even now: one must do it at night and keep the shutter open until lightning appears in the field of view of the camera. When the camera is fixed on a support like a tripod, everything is fine, but when one has no stable support, the camera will shake and move. In this case, street lights will create luminous traces on the film, but lightning is a very short phenomenon and will appear crisp. You can see this in the photo below (Fig. 3.1) which I took leaning against a pole: the lightning is quite sharp, but the street lights below leave jumbled traces because my hand holding the camera was shaking a lot.

The inexperienced photographers of around 1900 first interpreted these traces as tracks produced by moving ball lightning, and it was only after considerable debate that the correct interpretation was accepted. Unfortunately, books on ball lightning often waste a lot of effort on discussing this type of "evidence". The first photographs which are completely compatible with ball lightning are from an automated camera system called "Prairie meteorite network". This was a network of 16 cameras spread out over seven prairie states in the United States: every night they would automatically take photos of the sky in order to record traces of meteorites. The cameras had shutters closing periodically at 16 1/3 Hz, so tracks produced by meteors appeared as "dotted" lines, making it possible to measure the velocity of these objects. With more

© Springer Nature Switzerland AG 2019
H. Boerner, *Ball Lightning*, https://doi.org/10.1007/978-3-030-20783-0_3

Fig. 3.1 Photo of linear lightning with traces of street lights at the bottom. From the author

than one photo of such a track, triangulation was possible, and the end point of the track could be estimated. In almost 10 years of operation, only one meteorite was found by this method, so its efficiency was not really convincing. Nowadays, professionals[1] and also many amateurs[2] chase meteorites with much better electronic cameras, and they are also much more successful. For us the important thing is that the cameras took photos every night, regardless of the weather, so they also produced a considerable number of lightning photos. About 14,000 of these images were analyzed with about 120,000 lightning tracks (Tompkins and Rodney, 1980), looking for signs of ball lightning. The result was two candidates for ball lightning and 22 for bead lightning. Images of airplanes, cars, and firework rockets launched on Independence

---

[1] See https://fireballs.ndc.nasa.gov/
[2] http://www.meteorastronomie.ch

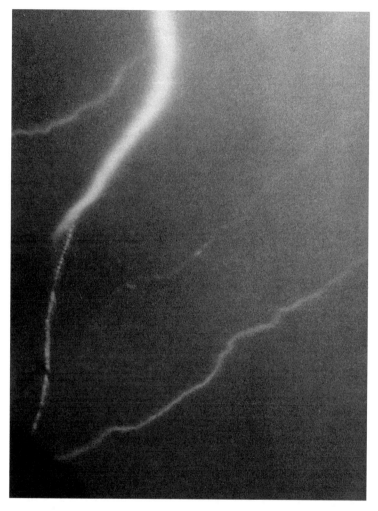

Fig. 3.2   Ball lightning photo from the Prairie network. From Stenhoff (1999)

Day were of course discarded. The best ball lightning candidate is shown in Fig. 3.2.

We can see three faint continuous traces of lightning in the background and one stronger continuous lightning trace in the foreground. From this trace, another trace exits at angle; it is not continuous but appears to be composed of individual points. The image requires a bit of explanation. Due to the oscillating shutter, it is like a composite of several frames of a video which are stacked upon each other. I made a drawing of the process in Fig. 3.3, showing how the consecutive exposures contribute to the final image. First,

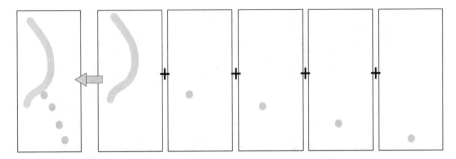

Fig. 3.3 Superposition of individual frames of the meteorite network photo. Drawing by the author

the linear lightning flash[3] was recorded, and the shutter closed. When the shutter opened again, a luminous round object appeared close to the lightning channel (it had lost its luminosity by then) and its progress down to the ground was recorded every time the shutter was opened again.

The authors assumed a height of about 2 km for the base of the clouds. Under this assumption, the luminous ball left the lightning channel at about 200 m height and fell to the ground at a speed of 62 m/s. Its diameter was between 2 and 4 m. The object on the second photo was further away and was not so clear as the one above. This type of ball lightning is different from the ones described in the reports above. It exited directly from the lightning channel and it was much larger and obviously contained more energy, because it was visible over a considerable distance. The question is obviously whether it was the same thing or whether we are now mixing up two completely different phenomena. Basically, we need to collect more information before we can tackle this question, but it will come up again and again in the book.

In their report, the authors discussed the possibility of confusion with bead lightning. So what is this about? Sometimes, when the trace of a lightning strike disappears, the channel is not extinguished uniformly: several regions may remain luminescent for longer, giving the appearance of a string of glowing pearls.[4] This phenomenon is fairly well documented. The first record of it was a movie taken of the explosion of an underwater test charge, which threw up a large column of water. The lightning hit the top of the column and decayed into several irregularly spaced bright regions (see Fig. 3.4). In the case of the photos from the Prairie network, this type of lightning would produce a trace which leaves the lightning stroke at the tip in the direction of the light-

---

[3] In fact, only part of the lightning discharged is seen in the image. Since the trace does not reach ground, it must be the leader and not the return stroke.

[4] I have once seen bead lightning, and the appearance was different from its usual description. See Case 24 in the appendix for my account of this observation.

Fig. 3.4   Photo of bead lightning. Film courtesy of S. Singer, scan by the author

ning trace, not at an angle as in the photo of Fig. 3.2. The authors reproduced one of the photos of such a bead lightning candidate, but the trace is faint, indeed almost imperceptible. It was not possible to obtain the original photo with better quality, so I cannot show it here.

These photos are certainly authentic and clearly show an object which falls into the ball lightning category, but authors like Mark Stenhoff have been more cautious (Stenhoff, 1999). Certainly, the photo in Fig. 3.4 needs some explanation, but it was taken by an automatic system operated by scientists; the authors have taken great care to eliminate possible misidentification and the image is consistent with other accounts. Due to the oscillating shutter, it contains more information than a normal photo, but less than a video or a movie, which would have recorded the images separately. Next, we will see how its looks when a photo is taken of ball lightning in the same way as in Fig. 3.4, leaving the camera shutter open for some time.

This photo was taken by Ern Mainka near Melbourne, Australia, in 2002. Ern was an experienced photographer, well known for his images of the National Parks all over Australia, but especially in Victoria. He is credited with saving several gorgeous mountain forests from destruction by timber companies due to the public impact he created with his beautiful images. On 1 February 2002, he was out taking photos of a violent thunderstorm near Melbourne. What then happened is best explained in his own words:

"The ball lightning was captured toward the end of a time exposure of a few minutes duration. This was the first of five consecutive frames taken of this particular storm cell. I also observed it visually (not through the camera lens) from the time it reached full intensity to when it descended out of view. The moment before it appeared, I was watching more intense strikes slightly to the right (see Fig. 3.5) and I cannot specifically recall noticing the strike that it appears to be connected with. It appears that this strike occurred before the ball lightning appeared and may perhaps be unrelated. Scientific opinion also suggests this to be the case. [...] However, at other times it has been observed to be 'striking off' cloud to ground strikes. I have seen BL perhaps three times before in Melbourne. Twice out of those three events I saw it striking off normal CG strikes. [...] It seems it can occur in either circumstances." The full account of his observation and the ensuing investigation is Case 2 in the appendix. The photo shows a linear lightning strike from cloud to ground, and near ground, a red trace appears to leave the lightning channel at an angle and follow in a curved path to earth.

You can see an enlarged part of the image in Fig. 3.6. At first, it is faint, but its intensity grows and develops a bright yellow core with a red envelope. Since the object was more than 20 km away from Ern, its brightness must have been considerable. Ern managed to find several witnesses who had probably seen the object and collected their testimonies; his summary is also given in the appendix (Case 2). It seems that the object starts from the lightning channel or very close to it, since an accidental superposition is very unlikely. It is hard to find an alternative explanation for this photo. An airplane appears impossible because of the shape of the trace, and a manipulation seems

Fig. 3.5 Ern Mainka's photo of a linear lightning stroke with a red ball lightning trace branching off (in the center). Public domain, from the State Library Victoria, Melbourne

Fig. 3.6 The ball lightning trace enlarged. Public domain, from the State Library Victoria, Melbourne

unlikely; the image was taken with an analog transparency film and shows the grains of the emulsion. Moreover, this is the twin of the object in the Prairie Network photo. Ball lightning like this is very rare, otherwise we would have more photos like this with so many people taking videos and photos of lightning with their cameras and smartphones. The authors of the Prairie Network investigation had only two clear candidates for 120,000 lightning traces.

Ball lightning with this energy content will obviously be visible over large distances, so it shouldn't come as a surprise that photos like the ones shown above are the first to become available. This is also true for videos, of course, as we shall see now.

This video was produced by a team of observers from the "Laboratoire de recherche sur la foudre", a lightning research laboratory in France. It was taken during a thunderstorm on 12 June 2003. Towards the end of the storm, which had produced mostly lightning in or between the clouds and only infrequent strikes to the ground, in the space of just 20 minutes or so, 21 very strong strikes called "super bolts" were observed to emerge from the anvil of the storm cloud. The sixteenth strike of this series produced an intense luminous object which was almost spherical. The object persisted even when the light from the lightning channel had vanished, surviving for a total of

1.4 seconds. Initially, its extent was 12–16 m, and shortly afterwards the halo was 4–5 m in diameter. The ball was not created at the exact spot where the lightning hit the ground, but—judging from the image—at least 10 m away. It didn't move throughout its existence. The researchers visited the spot where the lightning had hit ground, which was on the top of a hill with very little vegetation. No damage or other trace of either the lightning or the ball were found. More importantly, no man-made structures like power lines, transformers, etc., were found there. Very often discharges from power lines are mistaken for ball lightning, but this alternative explanation can be excluded in this case (Figs. 3.7 and 3.8).

A considerable number of ball lightning reports concern similar cases: bright spheres appearing close to, but not actually at, the point of impact of a linear lightning strike. In fact, more than 50% of some collections of ball lightning reports concern such cases.

The following report was published in a well-respected scientific journal, the Physical Review Letters (Cen et al. 2014), creating quite a stir in scientific

Fig. 3.7 Ball lightning near a linear lightning trace in France. With kind permission of R. Piccoli, Copyright R. Piccoli, Laboratoire sur la Foudre

Fig. 3.8 Later video frame, when the linear lightning trace was almost gone. With kind permission of R. Piccoli, Copyright R. Piccoli, Laboratoire sur la Foudre

circles. Such a recording had never yet been achieved and its impact was therefore considerable.

In the summer of 2012, a group of researchers were investigating lightning spectra at the Qinghai Plateau in China, using a so-called slitless spectrograph. This is basically a camera with a diffraction grating in front of the lens. It works well with lightning at night, because the thin lightning trace acts like the slit of a normal spectrograph, giving a nice, well resolved spectrum of the hot plasma in the lightning channel. The researchers used a 50 frames/s color camera and a high-speed monochrome camera of 3000 frames/s, both with diffraction gratings. During the night of 23 July, a normal cloud-to-ground lightning strike created a luminous object at the point where it struck ground. This round object was recorded with the color camera from start to end, whence its spectrum was obtained, and partially also with the high-speed camera. The total duration was 1.64 seconds, much longer than a normal lightning strike, which would be over in a few milliseconds. The object changed color from purple at the beginning, to orange, white, and red at the end. It moved at a speed of about 8.6 m/s in a horizontal direction. The lines in the spectra are characteristic of excited atoms or molecules and can be used to identify the composition of the hot gas and also its temperature. This spectral analysis allows us to investigate the composition of all luminous objects from a distance, the most important example being the chemistry and surface temperature of stars and other celestial objects. In this case, the lightning spectrum is typical of air at a temperature of 30,000 K. The lines are due to the spectra of nitrogen and oxygen, even due to single ionized nitrogen, i.e., having lost one of its electrons due to the extreme temperature in the lightning channel. The spectrum of the round object is different: emission lines from silicon, iron, and calcium are present and only weak lines from oxygen and nitrogen. Silicon, iron, and calcium are of course elements which are found in the soil, so the conclusion is that the lightning evaporated a bit of dirt where it hit the ground and that these gases were incorporated in the luminous object. Its temperature could not be measured exactly, but it must have been considerably cooler than the lightning channel.

The intensity didn't vary much over the lifetime of the object, but a strange thing was recorded by the high-speed monochrome camera: the intensity oscillated with a frequency of 100 Hz. At first this seems somewhat surprising, but with a little thought the source of this regular variation becomes quite clear: there was a 35 kV power line only about 20 m away from the spot where the lightning hit the earth. In China, the alternating current of power lines has a frequency of 50 Hz, and the power dissipated by an AC current of this frequency goes with double the frequency, which is 100 Hz. So how could such an object be influenced by the power line? There are two possible scenarios:

- If the object was indeed ball lightning, it would clearly contain free electrical charges like electrons and ions. These could be affected by the electric field of the power lines even at a distance, moving the charges in the ball lightning, especially the free electrons, and thereby heating the object.
- The object was not ball lightning at all, but a discharge at the power line induced by the lightning strike. In this case the variation in intensity was only the variation in the emission of the electrical arc.

Can we distinguish between these two cases? Unfortunately, this is not easy based on the information in this paper alone. For example, one would like to know if the power line was made of copper or aluminum, because these materials should show up in the spectrum. Copper would give a nice green color, but aluminum is not easy to identify in such a spectrum because its characteristic emission lines would fall outside the spectral range of the equipment used. No such lines are seen in the spectrum of the ball, but the discharge could also have happened at an insulator on the line, and this could have been covered with dust from the nearby soil. In this case the spectrum would be almost identical with the observed spectrum. Another point in favor of the ball lightning scenario may be the low temperature of the object; one would expect an electrical arc to be of very high temperature. In conclusion, since we cannot be sure how to decide this, I suggest that we put this interesting observation into the category "possible ball lightning, with somewhat unclear circumstances that need clarification".

You may ask why I have not mentioned any video published on YouTube, for example. There are quite a number of videos claiming to show ball lightning, and some of them may indeed be genuine, but these videos often suffer from one of the following problems:

- the images are blurred and out of focus, so it is difficult to judge what is actually shown;
- the objects or effects shown are clearly something else, like flash-over on HV lines or similar effects;
- the author claims to be an enthusiast of Ufology or other esoteric ideas, so the video might have been manipulated to create an especially interesting observation.

Summary

- There are very few still photos of BL which can be considered genuine, but the following two are reliable: one from the Prairie Meteorite Network and one from Ern Mainka, a nature photographer.
- Videos of BL offer better information on the creation of these objects, but there is some difficulty establishing whether they are genuine or fakes.
- I selected a video from France and one from a publication as examples.
- There is a growing number of BL videos to be found on the web, but as explained above, in most cases it is impossible to be sure that they are real.
- All photos and videos of BL that have surfaced so far are from objects which occurred at a considerable distance. Videos of BL objects in rooms, taken at close quarters, are still lacking.

4

A Bit of Philosophy, or What Has a Razor to Do with Ball Lightning?

The reader may now be wondering what the research activities on BL are under way within the scientific community, especially among physicists. From the information above one would of course expect a phenomenon with such a wealth of interesting properties to be the subject of intense study in several universities and other research institutions worldwide. In reality, just the opposite is true. Only a handful of scientists are publishing papers on the subject, and most of them disagree on the nature of BL. There are still fewer "mainstream" scientists working on the it. Even though practically all physicists have heard about BL, most of them seem to regard this subject either as intractable or unworthy of closer study. Quite a few scientists regard it as a pseudo-phenomenon, something that does not exist at all in reality (Campbell, 2008). I am not aware of any research project devoted exclusively to the study of BL. In order to show you where this lack of interest or even deep skepticism stems from, let us take a look at scientific investigation in general, starting with the three foundations that are essential to the whole of modern natural science.

Scientific Investigation and the Problem with BL

Over the last 300–400 years scientists have developed an elaborate system to separate fact from fiction. This system of methods is now in use in natural sciences like physics, chemistry, and biology, helping to recognize reality and discriminate against illusion, hallucination, and pure fiction.

The three foundations that support all the natural sciences are:

© Springer Nature Switzerland AG 2019
H. Boerner, *Ball Lightning*, https://doi.org/10.1007/978-3-030-20783-0_4

- controlled observation,
- repeatable experiments in different laboratories,
- theory and simulation.

Controlled observation is the starting point of all scientific investigation. This can be in the form of observation of nature, or from something that happens unexpectedly in the laboratory. The word 'controlled' means that today we would expect a recording by a device that is independent of the human experimenter, for example, a camera or a measuring instrument. This was of course not possible until the middle of the nineteenth century; scientists had to write down the effects they observed with their own senses: eyes, ears, touch, and sometimes taste. Photography was the first technological advancement which was able to provide an objective record of certain situations. Astronomers were among the first to take advantage of this, exposing large glass plates covered with photographic emulsion in order to record the position and brightness of stars and other celestial phenomena.

An observation is therefore expected to deliver data that are independent of the human observer in order to exclude individual peculiarities relating to eyesight or difficulties in recognizing or memorizing things. Criminal investigators know very well how difficult it is to get reliable information from witnesses. Some people are good observers, but most are not. Judgment of the sizes of objects, spatial distances, and especially time differences is fraught with error. In addition, we now know that our senses are rather limited. We can only see a small part of the spectrum of light waves, and we are blind to both higher energy light like ultraviolet rays and lower energy light like thermal radiation or infrared. For other things like electric fields, magnetic fields, radioactivity, to name but a few, we have no senses at all. For these reasons, wherever possible, scientists rely now only on objective data collected by suitably constructed instruments.

The phenomena observed in nature are then repeated in the laboratory in order to be able to control all environmental aspects that might influence the outcome of such experiments. These experiments must often be refined to exclude certain influences or to increase their accuracy. Repeatability of experiments is essential. This means that the results of different experiments designed to measure the same quantity must be identical to within the experimental error. There have often been claims of exceptional results or observations that have created a real stir in the scientific community, but then experiments at other labs designed to duplicate the results have failed completely. One good example is cold fusion, which turned out to be quite impossible to repeat.

Ideally, the last step is to construct a theory from the observational data and the results of experiments. This usually means a mathematical formulation of the essentials of the theory, allowing accurate calculation of expected observational data. A theory should be able to explain both observation and experiment, but it is also essential that a theory should make predictions of effects not yet observed. A brilliant example is of course Einstein's theory of gravity, the general theory of relativity. It explained the observed deviation in the path of the planet Mercury (the larger than expected precession of the perihelion), but also the prediction that the path of light could be bent due to the influence of masses, especially large masses. The latter was soon confirmed by observation, during a total solar eclipse, of the deflection of light from stars lying close to the limb of the Sun. The last triumph of the theory was the detection of gravitational waves produced by the collision of two black holes, and likewise by the collision of two neutron stars (Abbott et al., 2017).

It should now be easier to see why it has been so difficult for BL to become a focus of scientific study. BL is relatively rare, and both the time and the place of occurrence are unpredictable. In most cases, it exists only for a few seconds and it is rather small and not so bright. Most often, BL objects are visible only over short distances, and only rarely are they so large and luminous that they can be seen and recorded over greater distances. As I have explained above, we have only a few reliable photos of BL, and several possible BL videos, but none of these videos can be regarded as a watertight case. Videos or photos of smaller BLs taken at shorter distances are lacking altogether. For investigation of these objects we must rely exclusively on observer reports, something that is nowadays completely unfamiliar to scientists and tends to rouse their suspicions. Such suspicion is of course well founded in many cases, but when the observers are reliable individuals, and when their accounts are similar, pointing to a common cause, scientists would be well advised to take their reports a bit more seriously. There are some famous cases where scientists have dismissed certain observations for many years, only to find out that the observers had in fact been perfectly reliable, and that science had thereby ignored some important phenomenon for a considerable length of time. The first example concerns certain luminous phenomena in the upper atmosphere, like sprites, elves, and blue jets (which will be described in some detail later). Pilots of jet aircraft had consistently reported these brief flashes of light high above thunderstorm clouds for years, but no scientist had ever been inclined to believe them, even though such phenomena had previously been predicted by physicists. Only after a low-light camera accidentally caught such a discharge on video did investigations begin, leading to a wealth of information and completely new branches of lightning research. Now, even amateurs are catching these discharges on video or in photos.[1]

Another example is rogue waves. For many years mariners have reported huge waves, much larger than normal waves. Since the theory of water waves did not allow for such large waves, scientists simply did not believe these reports. Given the reputation sailors have of telling incredible yarns, they might perhaps be forgiven. But when a sensor measuring the wave height on an oil platform in the North Sea recorded such a monster wave, science accepted the fact that the reports were genuine. Today we know that these monster waves are not rare at all and that they may be responsible for a considerable number of shipping accidents. The theory of these monster waves is still not established in all aspects because complicated nonlinear phenomena are involved.

In both cases it would have been better for science to take a closer look at the reports and think about how one could establish or disprove them.

So, there is a lack of hard evidence for BL observations, but as Alexander Keul, an Austrian BL researcher, has shown conclusively (Keul and Stummer, 2002), the reports stretching back hundreds of years are remarkably consistent, pointing to what he calls a "core phenomenon" which has been described in more or less identical form throughout this long time span. At the beginning of this book, I gave you the accounts of some of my friends and relatives that have seen BL, and I am sure that these reports were not pure fiction.

The second pillar of scientific investigation is the repeatability of experiments. On many occasions, people have tried to produced BL in a number of different experiments, ranging from high voltage and microwave discharges to high current discharges from old submarine batteries (Stenhoff, 1999) and low-temperature burning of inflammable gases (Stenhoff, 1999). In most cases, some luminous objects have indeed been produced, and they vaguely resemble the luminescent balls of BL, but it has been impossible so far to create objects that duplicate all the reported properties of BL (Rakov and Uman, 2003). The latest and best investigated attempts were discharges in water-filled vessels creating luminous "plasmoids" (Versteegh et al., 2008), but these are quite obviously rising masses of hot and luminous air, partly in the form of a plasma, but apart from the luminosity and the roundish shape, they do not exhibit the properties of BL described above. All these objects have some but not all properties of BL objects, so it is probably best to call them "pseudo-BL".

---

[1] For example in http://www.meteorastronomie.ch/ergebnisse_tle.html

Most of the activity around BL has been focused on the development of theoretical models. Obviously, it is much cheaper to sit down, take a pencil, and do some calculations than to design, build, and perform experiments, so there are a huge number of different models of BL floating around. I have not counted how many, but there could easily be more than two hundred. This large number alone shows that it is difficult to constrain the development of theories using the observational facts, especially when the authors of BL theories choose to ignore certain aspects of the observations, as we shall see later. Basically, all theories struggle to explain the stability of such an object, its energy content, the visual emission from its surface, and its shape. In fact, the situation is even much worse: by testing the models against a few well-established facts almost all fail, and it is thus possible to falsify them.

It should now be pretty obvious why science regards BL as an oddity at best: the steps that would constitute a scientific investigation are unable to fix upon phenomena of this type; they simply fall through the established net of procedures.

But are there any other phenomena with similar problems? In fact, there a few observations presenting science with similar, although not quite identical problems. These include gamma ray bursts from space and radio bursts observed with radio-telescopes. During the 1960s, both the US and the USSR concluded that it was irresponsible to contaminate the atmosphere with radioactive debris by pursuing nuclear bomb tests in the open air (and they were certain that they could learn enough from underground tests), so in 1963 they signed a treaty banning nuclear testing in the atmosphere, in outer space, and under water. To check compliance with this ban, satellites with gamma-ray sensors were launched, able to detect the gamma radiation pulses from nuclear explosions on Earth. In 1967 one of these satellites detected a pulse of gamma rays which was clearly different from what was expected for a nuclear explosion on Earth, but its origin remained a mystery. It soon became obvious that they actually came from space and not from Earth. Over the next 20 years, many more of these bursts were recorded, but since their position and the time of their appearance was random and therefore unpredictable, little more could be learned except that their locations were evenly distributed over the sky. Things only changed in 1997 when a specially designed satellite could pin-point the position of one of these bursts with enough accuracy for optical and radio telescopes to be able to locate its afterglow. It turned out that it was produced by a very powerful event in a distant galaxy. Today more satellites are available to locate the position of these bursts very quickly. Then several telescopes in suitable places can swing into action to observe them. Very recently, such a gamma-ray burst was observed by these gamma-ray satellites and simultaneously by the first gravitational wave telescopes (Abbott et al., 2017). As soon as possible, almost all available telescopes swung into action to observe the event; about one third of the available astronomical observational capacity on Earth was studying the position of the burst. The results went beyond astronomers' wildest dreams: the gravitational waves clearly indicated that the burst was due to the cataclysmic merger of two orbiting neutron stars. The data from optical and radio telescopes produced a wealth of information on the nature of the collision and the properties of these fascinating objects. It took science exactly half a century from the first detection of these bursts until a firm conclusion could be drawn about their origin, and the effort involved in getting there was truly immense.

Recently, radio telescopes have observed very powerful millisecond-long flashes of radio radiation which so far cannot be explained. Again, as in the case of gamma-ray bursts, time and position are random and unpredictable.

For both types of bursts, the main observational problem is clearly this unpredictability of time and position, exactly as for BL, but the big advantage is that they are visible globally. The situation is also similar for discharges in the upper atmosphere, like sprites, which can be recorded over distances of hundreds of kilometers or even from space. Monster waves are now detected by radar satellites that monitor the whole surface of the oceans. BL in contrast is a local phenomenon, visible only over short distances, and it is much harder to envisage a surveillance program that could record their properties in nature. Only in the rare cases where BL is produced high in the air from a lightning channel or in a thunderstorm cloud is it actually visible over distances of several kilometers, and it should be no surprise that photographs and videos of BL have so far shown only those objects. The smaller BL objects seen indoors or at close range have not yet been recorded, as far as I know (This is also the reason why a photo of a plasma globe had to be chosen for the book cover, its appearance probably comes close to that of a real BL).

In addition, it seems that the physics of BL is not an easy problem, otherwise a good theory would already have been found. And the production of BL obviously involves conditions that are sometimes met by lightning, but not by any other setup that has so far been available in laboratories around the world.

Other phenomena have in the past presented hard theoretical problems: superconductivity was detected in 1910, but the correct theory was only developed in 1964. In 1986, high temperature superconductivity (high temperature means above the temperature of liquid nitrogen, so it's not really high compared with room temperature) was observed for the first time, but so far there has been no generally agreed theory on its origin, and predictions about the maximum temperature at which we can produce superconductivity are still lacking.

When analyzing the many reports on BL, we are now faced with a hard problem: the signal is very noisy, and we have no idea if it comes from just one source or if we are mixing signals from several origins. To draw an analogy, we have a pile of puzzle pieces (the observational reports) and we are not sure whether they are from just one puzzle or from several unrelated puzzles. In addition, many of the pieces are damaged and faded, so we cannot be sure what they actually depict. And just to make the problem harder, some people have smuggled fake pieces among the real ones. We clearly need some acceptable way to handle such a situation.

Wielding Occam's Razor

When scientists are investigating observations or trying to formulate theories, they are frequently faced with the problem of how to choose between different possibilities. This dilemma turns up constantly in various situations, but fortunately help comes from the teachings of a Franciscan friar living in England at the end of the thirteenth century: William of Occam (or Ockham). He lectured for some years in Oxford. He was of course interested, not in natural science as we know it today, but in theological questions concerning the bible. He constantly used a method in his discussions which has later turned out to be a very useful guiding principle in natural science and engineering. Translated from Latin to modern English, it states: "Entities are not to be multiplied without necessity".

This means that we should always chose the simplest explanation which is compatible with our observations. Additional assumptions should only be made if they are absolutely necessary, otherwise they are superfluous and should be avoided. In other words, one should not introduce new or additional concepts to explain observations unless necessary, even if it may be tempting to do so. As a corollary, one should always try to explain as much as possible of the observations in terms of already known scientific concepts. In engineering, a similar very useful design guide is called the KISS principle: Keep It Simple, Stupid. Simpler explanations, designs, or theories are easier to explain, understand, debug, verify, or falsify, and this can explain to some extent the success of this principle during the development of the natural sciences (although philosopers are puzzled why this principle appears to work so well).

Rather fancifully, Occam's principle is called a "razor", because one imagines it being used to shave away unnecessary parts of an interpretation of observations or of a theory. We shall see later on how it can be used to choose between different scenarios for the interpretation of BL reports. Since the science of BL is not very advanced and we have to deal mainly with observer reports (which may be only partially correct), great care must be exercised in handling the available information. Unfortunately, many people including scientists seem to apply different versions of Occam's razor when dealing with BL observations and theories. They sometimes seem to apply just the opposite, i.e., they choose the most fanciful interpretation. We will soon consider some examples of the use of Occam's razor, and some examples of its opposite version.

Summary

- The established procedure of scientific investigation rests on three criteria: well-recorded observations, controlled experiments, and theory plus modeling.
- BL partially fails the first criterion owing to its unpredictable and random occurrences, and it completely fails the second, because there have been no successful experiments so far.
- Most scientists think BL fails the third criterion as well, since they consider that there are no viable theories, but we shall see that this is not quite true.
- Other events which occur randomly in space and time are gamma-ray bursts and radio bursts, but they are observable worldwide and can be recorded with suitable equipment.
- Occam's razor is a powerful method for eliminating overly complicated or unrealistic hypotheses from scientific investigations.
- The careful use of Occam's razor is essential in analyzing BL observations, in order to establish its properties in a reasonable way.

5

Organizing and Analyzing
the Observations

Early Attempts

In the eighteenth century, considerable progress was made in understanding lightning and electricity. One of Benjamin Franklin's experiments is a famous example of the investigations done at that time: he flew a kite under thunderstorm conditions and was able to extract sparks from the wet, conducting line, demonstrating that lightning is electrical in nature. At the same time, machines were constructed that generated electricity by rubbing rapidly rotating glass balls, cylinders, or disks. These devices produced high voltages, but only low currents. Then Alessandro Volta detected the principle of batteries, opening the way to experiments with low voltages but higher currents. All these experimental results and inventions then led in the nineteenth century to scientific breakthroughs in understanding electricity and magnetism, culminating in the theory of electromagnetism due to James Clerk Maxwell in 1865. In other fields there was also much progress: it was understood that stones really could fall from the sky, and this led to an explanation of the fiery bolides that were sometimes seen streaking across the heavens, and the Solar System was extended with a new planet, Neptune, which had been found using predictions based on perturbations in the orbit of the newly found planet Uranus. It is obvious that, in this climate of general scientific progress, reports of ball lightning would also become the subject of scientific study. The French scientist François Arago was the first to publish an analysis of the reports available to him in his book "Sur la tonnerre" in 1838, but he concluded that it was one of the most inexplicable problems of physics at that time. Unfortunately, this is still the case almost 180 years on. In the second part of the nineteenth

© Springer Nature Switzerland AG 2019
H. Boerner, *Ball Lightning*, https://doi.org/10.1007/978-3-030-20783-0_5

century, small compilations of ball lightning observations were produced by several people, but the first serious effort to organize and analyze the data was only made after WWI by the German teacher W. Brand.

Brand's Book of 1923

Brand's intention was to collect reliable and sufficiently detailed reports on BL to be able to extract information on the observed objects and their properties.

There are several reasons why his book (Brand, 1923),[1] stands out from the few earlier attempts to collect BL reports. One is that he took a very critical look at the information contained in the earlier compilations, since he found quite a number of errors in the second-hand reports. He therefore tried to go back to original reports whenever possible, to exclude transcription errors and exaggerations. In one case, the original source mentioned the duration of the BL event to be a few seconds, while in the secondary report it had turned into minutes. Furthermore, he checked the reports and omitted anything that contained insufficient information and anything that he considered to be insufficiently credible. He never mentioned it by name, but he clearly applied Occam's razor on several occasions. One good example is his handling of cases in which the BL was supposed to have made a metal object to "disappear" without trace, especially gold objects. Instead of this somewhat fanciful interpretation, he suggested that the victims of a normal lightning strike had been robbed of the valuable objects while they were unconscious, this being a much more realistic hypothesis. This is an excellent example of how Occam's razor should be used; Brand would probably have called it "common sense". To stay with our puzzle analogy, he discarded all the pieces that were damaged or faded and tried to identify and discard fake pieces that had been smuggled into the set of genuine ones.

He further identified two major sources of confusion between BL and other phenomena: first normal linear lightning and second meteors, and he omitted the corresponding reports. In this filtering process he reduced the number from about 600 cases to only 215, which he then subjected to a thorough analysis. His book is organized in three chapters. First, he explains his method and the reason why he has excluded certain reports from his analysis. The second part contains references to all 215 reports and the full text for the most important ones. The third chapter analyzes the various cases, always providing a reference to the report that has led him to his conclusions. The chapter

---

[1] English translation: (Brand, 1971), recent reprint: (Brand, 2010).

closes with a brief overview of ball lightning theories existing at that time. Initially, his book was not very influential, because it was only available in German until 1971. In this year NASA produced an English translation, but it is difficult to obtain (Brand, 1971). Moreover, even the initial circulation of the book may not have been large since the publisher complains in the foreword that he had great difficulty in financing the publication, given that the book was published at the height of the period of hyperinflation (1923) in Germany. Recently, a new facsimile edition has become available in Germany (Brand and Wittmann, 2010) through the efforts of Axel Wittmann, an astrophysicist who witnessed a spectacular case of ball lightning when he was a child (see Case 9).

Even today, most of Brand's conclusions are still valid, and indeed, not so many new facts on ball lightning have been added since 1923.

An important result which he established was that BL is usually, although not always, appearing close to (but not directly at) the strike point of normal lightning which probably initiates the BL object. He also stated that winter thunderstorms are relatively richer (per lightning stroke) in BL objects than summer thunderstorms. We will come back to this surprising correlation later, since it was the starting point of an investigation I have been working on for several years. For the dissipation of BL, he found three modes: silent disappearance, explosive disappearance, and being hit by a second normal stroke of linear lightning. We will discuss this in more detail in the section on BL energy below.

Brand's attempts to understand the nature of these objects was of course shaped by the state of physics in his day. In 1923, quantum theory had not yet been developed, the nature of the atomic nucleus was still obscure, and the term "plasma physics" had not been coined. He therefore had to rely on the limited amount of knowledge about electrical discharges and lightning existing in those days.

Forty-eight years, almost half a century, were to pass before the next book on ball lightning was published.

Singer's and Barry's Books (1971 and 1980)

Singer's book "The Nature of Ball Lightning" (Singer, 1971) was the first book on the subject published in English. Its style is different from Brand's book, which was an analysis of the observations. In contrast, Singers intention was to give a complete overview of the subject in form of a review, not presenting his own work on ball lightning. He repeats Brand's list of deduced properties and compares it to several compilations that had been performed before 1971. A large part of the book focuses on ball lightning theories, which had progressed considerable since Brand's book. Singer's conclusion was that ball lightning is a form of plasma object, but he acknowledged the difficulties inherent in this interpretation. Barry's book (Barry, 1980) resembles Singer's, but he concentrated more on experiments to recreate ball lightning objects in the lab. His favorites were experiments with dilute flammable gases, but he did not explain how this was compatible with the fact that ball lightning is observed in conjunction with lightning.

In both books, a sizable part is spent on the discussion of genuine or purported ball lightning photos.

Both books are very readable, especially the parts on observations and deduced properties, but the chapters on theory and experiment are rather out of date.

Stenhoff's Book (1999)

"Ball Lightning—An Unsolved Problem in Atmospheric Physics" is the best book on the subject so far (Stenhoff, 1999). Stenhoff had done quite extensive research on several cases reported in the UK and he covers almost all aspects of ball lightning. He was also able to draw information from the progress that had been made in understanding thunderstorms and normal lightning. This is the first book on BL that contains a glossary of terms used in lightning research.

To the best of my knowledge, he was also the first to take seriously Brand's conclusion that winter thunderstorms are richer in BL than summer thunderstorms, and he also gave a tentative explanation for this. This will become quite important in our attempts to understand the nature of BL.

Books from Russian Researchers

Researchers from Russia have been quite active in BL research since the 1950s, but unfortunately, a significant number of these publications, especially the earlier ones, are only available in Russian. This is also the case for Stakanov's books from 1979 and 1985 ("The Physical Nature of Ball Lightning" and "About the Physical Nature of Ball Lightning"), where he presents a huge collection of observational reports (more than 1000 cases, according to Smirnov, 1987).

Compilations of BL Observations

As explained above, Brand produced the first serious compilation of BL cases by filtering the reports in newspapers and magazines according to information content and reliability. The next compilation of observations was made by Humphreys in the 1920s and 1930s. He collected 280 reports and published a short paper on these in 1936 (Humphreys, 1936), but unfortunately he was only interested in demonstrating that almost all of them were due to already known phenomena. He did not make a thorough analysis like Brand and did not give any statistical information. We will discuss his paper in the chapter on skeptic's views later. The next compilations were made in the 1960s. The renewed interest in BL was probably due to two factors. First, plasma physics, the science of ionized gaseous matter, had been developed in the meantime, and the research on controlled fusion of heavy hydrogen was an important topic at the time. In addition, at the height of the Cold War, there was a certain anxiety in the US that the Soviet Union was ahead in this area of research, and there was some research money available to investigate even more remote chances of making a breakthrough in this area. Two compilations of BL cases were made using a new technique: employees of two research centers were asked if they had seen BL and were then given questionnaires with multiple choice questions in order to obtain a set of reports with more structured information. The first survey was made by McNally, involving the complete personnel of Union Carbide Nuclear Company in the Oak Ridge complex, resulting in a total of 513 reports from 15,923 individuals (McNally, 1966). The reason for this report is clearly given as analyzing BL with respect to nuclear fusion research. He states: "Ball lightning, as a possible stable plasma configuration, is receiving increasing emphasis in research especially in the USSR." The results of the survey concerning the observed properties of BL are more or less identical to the results obtained by Brand.

A second report by Rayle (1966) was compiled from the responses of 1764 observers of lightning and ball or bead lightning. He concluded that "BL was not a rare phenomenon and that the frequent occurrence of ball lightning events demands that any explanation put forth to account for a significant fraction of such events not depend on extremely unlikely circumstances." I agree with him that there are more observers than one thinks, based on the experience with the reports from my friends and acquaintances. The circumstances for BL creation, on the other hand, must be very special, otherwise we would observe many more and we could reproduce them in the laboratory.

Rayle did not adopt the filtering approach used by Brand. He accepted as BL everything the observer identified as BL. Such an approach is risky. It can only work if the percentage of misidentification or even hoaxes contained in these compilations is small enough. Stenhoff found (Stenhoff, 1999) that this was not necessarily the case, as more than 80% of the cases in his collection could be explained by already known phenomena, such as St. Elmo's fire or normal lightning. His conclusion was that a few well documented and carefully investigated cases could be much more useful than the many reports in the statistical surveys. In principle, I support this view and I will give several examples below that in my opinion are due to real BL observations and are very well documented. I also ask whether the observations presented at the beginning of the book could have alternative explanations, and I am quite sure that this is not the case, so Stenhoff's collection may not have been so typical, and his conclusion may be a bit too harsh.

Since the 1970s, many more collections of BL sightings have been made. From Europe, Keul (2008) lists France with 249, Germany with 332, Austria with 253, Hungary with 281, and Russia with 3104 cases. In addition, there are collections from the US with 282 and Japan with 2060 cases (Ohtsuki and Ofuruton, 1991). The quality of these collections is rather uneven. Keul, who has been working on the subject for more than 40 years, applied a filter like Brand's, but there is not enough information available to judge the quality of the other compilations.

The Russian compilations come with very high numbers of reports, but only extracts have been published, or some analysis with respect to specific topics. Japan is a special case. There is a collection with many cases, but very few details have been published, and it appears that there may be differences with the BL characteristics derived from the other reports.

There are only very few reports from the Tropics (India), with Africa and Brazil missing completely. The global lightning hot-spots are in the Congo Basin and Venezuela, but there is no information at all on BL sightings from these regions.

There could be a number of possible reasons for this lack of reports. For example, there may be fewer BL objects occurring in these regions, there may be superstitions with respect to these mysterious objects and people may be reluctant to talk about them,[2] or people may just have more serious problems to deal with than reporting BL observations. But all in all, the reasons for this very uneven spatial distribution of reports is basically unknown. From Australia, on the other hand, BL reports are widely available.

---

[2] This information was given in a personal communication.

What can be done with this amount of information? The most recent compilations are due to Alexander Keul, an Austrian researcher who studied both meteorology and psychology. He carried out a very interesting study by comparing different compilations—two from Austria and one from Germany— with respect to the BL characteristics observed. His surprising finding was (Keul and Stummer, 2002): "Summing up the results, the three data profiles suggest one time-constant, space-invariant set of BL phenomena over Central Europe."

He also compared the Central European cases with reports from the US. His conclusion was: "Summing up, the detailed Beaty data bank compared to Central European data shows a very similar BL phenomenon with regard to observer mean age, reaction, number of observers, thunderstorm connection, object number, form, size maximum, motion, residue and simultaneous sound. BL data are different with respect to gender proportion, maximum years and months, daytime maximum, simultaneous precipitation and lightning flash, duration, distance, color peak, brightness and final explosion. Whereas the observer populations are highly similar, some weather conditions and especially some object data (duration, distance, color, brightness) are not."

Being a psychologist, he also investigated the observer characteristics to find out if there was any kind of bias.

Obviously, even when one considers that the BL compilations contain a certain amount of "noise" in the form of misinterpretation or plain fakes, a core phenomenon emerges from the reports.

Reported Characteristics of Ball Lightning

From the reports mentioned above one can make quite a comprehensive list of the observed properties of BL. The reviews agree on these properties, with the notable exception of the energy content of such objects.

- The occurrence of BL is very well correlated with thunderstorms, but there are exceptions. Several observers of BL were sure that there was no thunderstorm close by. In addition, Brand arrives at two interesting conclusions: winter thunderstorms are richer in BL than summer thunderstorms, and BL is more often seen at the end of thunderstorms. These observations offer very important clues on the physics of the way these objects are generated, and these clues will be discussed in detail below.
- Brand concludes: "Very often the ball lightning is preceded by an initial flash, whereby the ball lightning is formed at the site where the lightning

strikes or in the vicinity thereof; however, the absence of an initial lightning discharge is not a rare occurrence." The point of creation can also be at a considerable distance from the impact. A good example is the Neuruppin case (Appendix, Case 1), where at least 11 BL objects were created more than 5 km away from the impact, so a direct contact with the lightning channel is not necessary.

- The duration of the BL objects can be as short as a second, but they can last for minutes in some cases. At Neuruppin, some events lasted for just about a second, whereas others could be observed at leisure, suggesting that they lasted many seconds.
- It is obvious from their luminosity that BL objects emit energy, so they will store a certain amount of energy. The exact value must be deduced from interaction of the BL with its environment, so only a few cases have been reported. There is an intense debate about how large the energy content can be, so there will be a more detailed discussion below.
- In most cases the shape is spherical. Smirnov (1987) claims that about 90% of all observations report spherical BL; the remaining 10% are split between ellipsoids, toroids, pear-shaped structures, and irregular shapes. Sometimes the objects have a complicated internal structure, as in Case 6. In Neuruppin, ellipsoids were observed twice. A single large ball lightning object can burst into several smaller balls as in Case 9. It may happen, although very rarely, that two ball lightning objects appear to be connected by a "string" consisting of small luminous beads, or that a short row of beads occurs with a single ball lightning object (Case 2, observer 3).
- The color of BL objects can vary over almost the whole visible spectrum: colors from deep red, through orange-yellow to blue and blinding white have been reported. Only green is rarely seen.
- There is also a considerable variation in the surface structure. Some objects have a solid appearance (Case 19), whereas others are transparent as in Cases 6 and 12. Some objects appear to have a very active surface, throwing out sparks (Case 14) or with continual changes in their surface, whereas others present a quiet surface (Case 19). It seems probable that the objects that appear non-transparent just have a higher luminosity than the transparent ones, so any light from the background scenery is swamped (Stephan, 2012).
- Brand comments on the motion of these objects: "The speed of ball lightning is very great when appearing at the base of the cloud and shooting down to Earth [...] Near the Earth's surface or in closed spaces, ball lightning moves at a speed of approximately 2 m/s [...] At times the ball can be carried along with the wind, though its motion is usually independent of

the prevailing air currents." He continues: "They are 'attracted' by the air in closed spaces; these balls enter the latter via an open window or door, and even through narrow cracks; however, they display a marked preference for entering via the flue gases of chimneys which have a better electrical conductance without self-induction, so that very often they emerge from the hearth and thus gain entrance into the kitchen. After the ball describes a few circles around the room the latter seems to lose its force of attraction and the ball leaves again via an air passage, often the same as before, though at times along a new path." The motion of BL objects inside rooms is sometimes described as if the objects were animate. At Neuruppin (Case 1), the BL objects displayed many different kinds of motion: some were at rest during the observation, but others moved fast or made a round trip of the room. One moved towards the chimney, as in the old reports from Brand (Case 1, witness 4).

- BL objects have been observed passing through holes, curtains, and windows; sometimes even through metal screens. This behavior is very much the focus of a debate among BL researchers, so a detailed discussion is given below.
- The existence and creation of BL objects in closed rooms and in modern, all-metal aircraft has often been reported. This characteristic considerably restricts the possible theoretical models, so it is also discussed below.
- There are mainly two decay modes for BL objects: they can vanish silently, but they often decay with a sound like an explosion. In Case 3, a wave of heat is mentioned when the BL object terminates. Brand also concludes that BL seen outside is occasionally terminated by a streak of lightning which strikes the ball.

Observations of BL Creation

Most of the time BL is only noticed after its creation. The actual creation process has only rarely been witnessed. In Neuruppin, two objects were seen when they formed inside houses, (Case 1 and witnesses 16 and 20) appearing "out of thin air". There are similar accounts, as in Case 6, where the BL appeared over a stone that was being lifted on a road by four men, outdoors in Australia (Case 13), or Case 8 where it appeared indoors. The physicist Loeb observed the creation of a BL when he was a child (see Case 7). In all these cases, no lightning channel was directly involved.

BL and Winter Thunderstorms

Brand was the first to analyze annual trends in BL incidence. He compared the absolute as well as the incidence of BL relative to the incidence of thunderstorms for observations in Central Europe. There was a clear correlation between the frequency of occurrence of both phenomena, with a peak in the summer months from June to August, but the relative incidence of BL as compared to the number of strokes was lower in summer and higher in winter.

He concluded: "Accordingly, winter thunderstorms produce a higher incidence of ball lightning than spring and summer thunderstorms."

He also checked the diurnal trend based on 108 observations from Central Europe, where he also had data on diurnal thunderstorm trends. BL occurred at a lower incidence during morning (midnight to noon) thunderstorms than during afternoon (noon to midnight) thunderstorms. His conclusions were: "This seems to concur with the previously made observation that ball lightning occurs particularly often toward the end or during the final stage of thunderstorms, which, of course, takes place more frequently during the afternoon than in the morning hours."

Both observations were basically ignored by BL researchers until 1999. Since Brand's data set was not large, we must check whether his conclusions are supported by later BL data compilations. Curiously, not all of these provide information about the month of the observation. All mentioned that BL is mainly observed in the summer, correlating well with the maximum of lightning activity, but the breakdown to monthly numbers was missing from some of them. Obviously, the authors didn't think it important, otherwise they would have included that data. In order to arrive at enough numbers, I had to combine the data from October to March as the winter half year and data from April to September as the summer half year. Results are shown in Fig. 5.1. The errors given are just the statistical errors given by the square root of the case number. One must be a bit careful, because not all these data sets are independent; some data have been incorporated in several collections. For example, Smirnov also included the collections from McNally and Raille in his paper, but since the number of cases from Russia was far greater, the results will not have been much influenced (Smirnov, 1987).

The data sets from Brand covering Central Europe and from France come from geographically close regions, but I have also included data from North America and Russia which may have different characteristics (Boerner, 2016). For comparison, I used the publicly available data from the lightning location system BLIDs in Switzerland and Germany in the years 2000–2003 and from

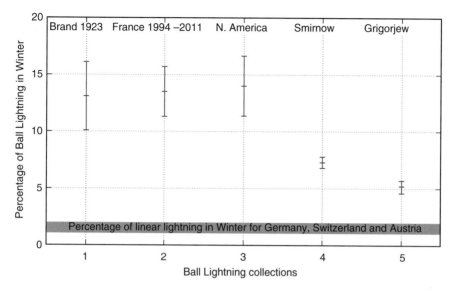

Fig. 5.1 Ball lightning percentage in the winter half-year for several BL compilations. Diagram by the author

the system ALDIS in Austria in the years 1992–2015; the percentage of winter lightning is indicated as the shaded band in the figure. For North America, no detailed lightning data were available to me. The lightning incidence for Russia is unclear since there is no lightning detection system covering this vast expanse of land. However, both North America and Russia have a mainly continental climate, so the data from Central Europe could be used as a rough guide.

Lightning in Central Europe is very strongly concentrated in the summer months, with only a meager 1–2% of the strokes in the winter half year. Of course, there is a variation from year to year, but it should average out since the time span was long, at least for the ALDIS data. BL observations follow the trend for all kinds of lightning to some extent, but instead of 1 or 2%, more than 7, sometimes up to 14% of the BL objects are reported in winter. The results clearly show a statistically significant excess of BL objects in winter thunderstorms, and this is true for Europe, North America, and Russia.

What could be the reason for this very surprising result? Two possible explanations come to mind: either BL objects are created with higher probability by the few winter lightning strokes than by the summer lightning strokes, or they are created with equal probability but are observed better in winter than in summer. It is hard to find a reason why people could observe BL better in winter, so we are left with the difference in lightning. Since BL is generally created by a normal initial lightning stroke, one could speculate that lightning in winter is different from lightning in summer and that this difference is the reason for this nontrivial property. This is indeed the case as we know now, but in Brand's time it was still unknown. Brand's finding was completely ignored until Mark Stenhoff made a link between this correlation and the properties of winter thunderstorms. Since his book was published only in 1999 and since most compilations were made earlier, there is simply no possibility that people could have fudged the results in the compilations. The second finding, the concentration of BL incidence towards the end of thunderstorms, also indicates a difference in the characteristics of the lightning involved, but this was not known in Brand's day either. We will learn about this in the chapter on lightning physics.

BL Energy Content and Energy Source

The luminous surface of BL clearly radiates energy away in the form of light. This luminosity could be due to several effects, but it is generally assumed that it is due to the radiation of excited atoms or molecules in a plasma. The question of where the energy for the luminosity comes from and how much energy is related to a BL object has been the topic of a considerable debate (Stenhoff, 1999; Bychkov et al., 2002). The energy could be stored in the BL when it is created and then emitted slowly, or it could be transmitted into it from outside more or less continuously during its existence. A combination of these extreme cases is also possible. Another question is the amount of energy that can be attributed to a BL. This is a tricky question, because this energy is not observable like the shape, color, motion, or duration. It must be deduced from other things like the brightness of the BL. Observers are often able to make a comparison of the light output of a BL object by comparing it to the light from an incandescent lamp. For example, Jennison (Case 19), who saw a BL at close range, compared its optical output to a 5–10 W lamp. Normal incandescent lamps, which work with a tungsten filament heated to about 2700 K, convert electrical energy into visible light with an efficiency of about 10 lm/W, so they convert only 5% of the electrical power into visible radiation. If we take this rather low efficiency value as a rough guide, then the total power (including all losses) of Jennison's BL would have been 5–10 W. The luminosity is often described as comparable to a 10–100 W light bulb, clearly observable in daylight, but not blinding. In a lifetime of say 10 s, the BL would then consume 100–1000 J of energy, of which 5–50 J would be visible light and the rest losses due to heating or other loss channels. With a diameter of 20 cm, the volume would be about 4000 cm^3, so the energy density would be 0.025–0.25 J/cm³. Stenhoff tentatively gives a slightly larger energy content of 3 kJ (Stenhoff, 1999). The energy content is of great importance when one tries to explain how BL objects can be formed in rooms inside houses, as observed for example in Neuruppin (Case 1, witnesses 16 and 20). Since many BL objects have an optical output that is below that of a 100 W lamp, they are not particularly powerful objects. It is of course questionable whether the efficiency of incandescent lamps has anything to do with the efficiency of BL in creating its optical emission. The tungsten filament of incandescent lamps is roughly a black-body radiator, which at this temperature emits mostly in the infrared region of the spectrum (see Fig. 5.2). Since no heat sensation has been reported from BL objects passing people at close range, BL cannot emit this amount of infrared radiation, so it is obviously not a black-body radiator. On the other hand, it could easily lose additional energy by heating the air close to its surface, so this efficiency value may be not a bad guess after all. On the other hand, the efficiency could also be closer to that of modern fluorescent lamps, which is much higher, about 100 lm/W. In that case, the energy content and the energy density values given above would have to be scaled down by a factor of 10. The conclusion is that at least indoor BL objects are rather weak.

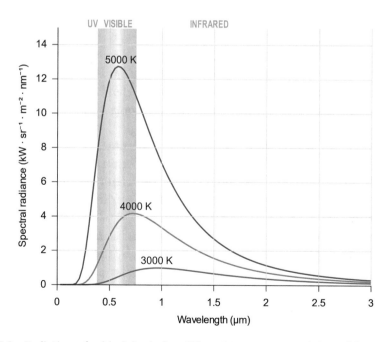

Fig. 5.2 Radiation of a black body for different temperatures. Adapted from: Darth Kule (https://commons.wikimedia.org/wiki/File:Black_body.svg), "Black body", drawing, public domain

Another way of reaching an estimate of the energy is based on the interactions of the BL with its surroundings. An often-cited example is the "BL in the tub" event of 1936 (Case 20), where a BL was said to enter and heat the water in a tub. Stenhoff was rather critical about the conclusions since the report is not as detailed as one would wish. The amount of heating is not clear, and it could also be due to ordinary lightning striking the tub. Other accounts have been collected which attribute a lot of damage to BL, but Brand already cautioned against naively using these reports, since the damage pattern was almost always compatible with the damage by ordinary lightning.

Stepanov (1990) compared the amount of energy of BL seen indoors to BL objects observed outdoors. He found that there is factor of 1000 between the energy estimates for the two groups. Indoors, the average energy was 100 J, whereas outdoors it was 100 kJ. He concluded that BL outdoors receives energy from an external source, which does not work for the objects seen indoors.

Brand already found that outdoor BL tends to be hit by a linear lightning stroke which then terminates the BL. The normal lightning then deposits its full energy in the vicinity of the place where the BL has last been seen. If the damage done by the last linear lightning stroke is then taken as a sign of the enormous energy of the rather small BL, this leads to extreme energy densities for BL which are often higher than the energy content of explosives like TNT. This is clearly an unrealistic hypothesis, so the explanation offered by Brand—the BL being hit by normal lightning—is to be preferred (Occam's razor again). Stenhoff comes to the same conclusion. If these considerations are taken into account, the number of remaining, credible BL reports indicating a high energy content is very small. We will come back to this point after we have looked at the physics of normal lightning.

In his report (Rayle, 1966) on the study of BL observations, W. D. Rayle remarks: "There is little indication that ball lightning commonly involves large quantities of energy. Very bright, noisy or destructive occurrences were few. A mechanism for ball lightning need not account for mega-joule energies to be satisfactory for the vast majority of cases."

The problem with the high energy content of BL is that it runs counter to one of the hard limits imposed by physics, in fact, by the so-called virial theorem. In order to store energy in a limited region of space such that the energy density is higher there than in the surrounding atmosphere, one needs a wall which can bear pressure, like the steel walls of a pressurized air bottle or—for plasma configurations—"walls" made by magnetic fields produced by coils.

Singer (2002) comments on this problem: "Possibly the major obstacles in the impasse are the somewhat long luminous life (a few seconds or more) reported in many observations of ball lightning and the energy displayed in a few cases exceeding the limitation set by the virial theorem. This rule limits the energy content of the fireball rather severely. The long lifetime implies a temporarily stable structure and radiating species existing or being generated during the globe's existence. The energy required for the observed luminosity in most cases need not be appreciable."

The virial theorem limits the energy content to perhaps three times the containment provided by the external surroundings of the ball, primarily atmospheric pressure. Stenhoff avoided the problem of the virial theorem limitation on ball lightning energy by ascribing all cases in which the ball appeared to release great energy to the effect of an accompanying stroke of common lightning. Nevertheless, there are some reliable BL observations that point to a very energetic object; for example, the BL observed and photographed by Ern Mainka was clearly emitting a lot of energy because it was observed over a distance of many kilometers.

Passage of BL Through Dielectric Objects

The passage of BL objects through closed windows is one of the characteristics that was and still is hotly debated among people interested in BL. The reason is that this feature restricts the theoretical models in a very fundamental way. The reports can be grouped the following categories:

- In a small number of reports, the BL object was directly observed to pass through a dielectric barrier like glass without difficulty. These are the reports that are least easily dismissed.
- More reports are available where the BL was only observed when it was inside the room; it was not seen entering or leaving through the window pane. There are only two possible explanations: either the BL passed through the pane or it was created directly behind it and also directly extinguished in front of the pane (when leaving), thus avoiding the passage.
- In several reports, damage to the window was reported. A hole was produced at the point where the object passed, most of the time an almost circular piece of glass was somehow cut out.
- There are also cases where BL passed through other objects, such as curtains.

The first report of a passage through a window is from Brand (Case 10 in the appendix). The BL passed through the upper part of a window of a hotel and vanished through a wall made of wooden panels, entering the neighboring room. The report is very clear; the observer visited the location again and gave a good description. It turns out that the hotel was quite well known, so a number of picture postcards of the building still exist.

Figure 5.3 shows the front of the hotel facing the street. There are three types of windows on this side of the building at street level. The BL must have passed through the upper part of one of them.

The next reports are from an event at the Cavendish Lab in 1982 (Case 14). During a thunderstorm of extreme intensity, several BL objects were observed in the lab. In one case, the BL passed through a window:

"[…] A closer encounter was experienced by an administrative assistant in a duplicating room on the ground floor. She was closing a window when she was startled by a noise that suggested the window had been knocked in. A bright, spinning, sparkling object of pyrotechnic appearance entered past her head, rebounded from a copying machine and departed as it had arrived. She said, 'It came in through the window, spinning, rolling, throwing out all sorts of sparks like a Catherine wheel. I was terrified.' The window was undamaged. Another person in the same room was convinced that something had entered the room".

A recent event in Devon (in fall 2017) is particularly well documented (Case 15). The observer notes: "It was a ball approx 30/40 cm diameter. I was

Fig. 5.3 Photo of the hotel where a ball lightning was seen passing through a window. Public domain

1.5 m from it. There was no smell, but the thunder noise was more or less at the same time and was very loud. The ball was very intense bright blue and was very sharp and you could not see through it. It came in a closed window and out through the glass of the closed patio doors; both window and doors were double glazed. The ball moved from left to right approximately 6 m and about 1.5 m in front of me at about 1.5 m high. I did not see the ball outside, but my neighbor did and also it turns out another lady saw it traverse the car park about 30/40 m and this was after it left me through the glass." The BL was initiated in this case by a negative cloud-to-ground ($-$CG) flash of moderate intensity (-15.4 kA as measured by EUKLID).

There are two reports of such passages in the book on lightning by Rakov and Uman (2003). In one case, the observer states very clearly how the BL entered the room through the window and blind (Case 17). In the other case, the BL passed a metal screen and a mica window, puncturing both (Case 16).

From Russia, there is a quite extensive report on similar cases. Grigor'ev published a compilation of cases where BL had penetrated windows (Grigor'ev et al., 1992), in some cases leaving the glass panes intact, in some cases, creating round holes.

In Neuruppin, several of the smaller BL objects were reported to have moved through curtains or—in one case—through a window (Case 1, witnesses 1, 4, and 25). In one case the cotton curtain was reported to have two brown spots at the position where the BL penetrated, while the other curtains and the window were undamaged.

These reports show that this kind of behavior is not so rare for BL objects.

This criterion is a real killer for most of the models. Glass windows are meant to keep out all physical objects like cold or hot air, rain, snow, insects, and so on, while allowing sunlight to pass unobstructed. So, the only things that can freely pass are:

- electric fields, static or time varying,
- magnetic fields,
- electromagnetic waves in the visible range, partially in the infrared and radio waves, including microwaves.

Glass is an excellent insulator so direct current cannot pass, but alternating current can, and the glass then acts like the dielectric of a capacitor.

Old windows were almost always made of a single pane of glass. In colder regions, double pane windows were used for better insulation, but the two panes were not hermetically sealed. Usually, the windows consisted of two frames that could be opened to clean the two panes. More recently,

Fig. 5.4 Window with a conductive coating giving a signal in a metal detector. With kind permission by W. Heil

double-glazed windows have been made with two hermetically sealed glass panes with dry gas in-between to prevent condensation of water. Initially, dry air was used for the filling, but now noble gases like argon or even krypton are used because they reduce the heat loss. Until recently, also, sulfur hexafluoride (SF_6) was used. It insulates well, and not only against heat but also electrically. It is a strong greenhouse gas, so its use in windows is now prohibited in Europe, but it is still heavily used in high voltage installations because it strongly suppresses electrical discharges. In noble gases, on the other hand, discharges are easily generated.

Modern windows often have glass coated with materials that reflect infrared radiation. This helps to keep rooms cool in summer, preventing the heat from coming in, and warm in winter, when the heat radiation needs to be kept in the house. For these almost transparent coatings, silver and metal oxides like tin oxide are used. Besides blocking the heat radiation, they also shield partially against radio waves because of their conductivity. It is easy to check whether the windows have glass with such a coating: in the figure below, a normal metal detector (usually for detecting pipes and electrical wires in walls before drilling a hole) is used on such a pane.[3] Uncoated glass panes do not give a signal since they are, of course, completely insulating. The coated windows absorb about 99.9% of the incoming radio waves, so one might expect them also to block BL objects like a metal plane (Fig. 5.4).

---

[3] The conductive coating is on the inner side of the double panes, so one cannot use an ohmmeter to check for the coating.

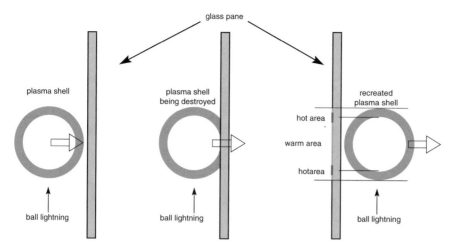

Fig. 5.5 Ball lightning with a hot plasma shell passing through a window. Drawing by the author

It is important to know precisely what kind of window is referred to in the reports of BL passages, but this is not possible for older reports, so one must guess what was most likely being used at that time and in that particular region. For recent reports, like the report from Devon, it was possible to get more information; here the double-paned windows were 15 years old. At that time, coated windows were not yet being used.

The reports on holes drilled into glass windows by BL objects have been much disputed. Skeptics claim that the holes were more likely produced by objects hitting the window, like footballs or similar round objects. From personal experience, I know that footballs are more prone to shatter a window pane into shards; they do not punch out round pieces of glass. Is it possible that BL objects could achieve such a feat? If we assume that BL objects consist of plasma, and that the whole sphere is filled by it, there is a problem: when hitting the pane, the ions and electrons of the plasma will recombine and heat the outer surface of the glass, but the heating will be greatest in the center of the circle where the BL hit. Such a heating pattern would probably break the window or break shards out of the hole rather than neatly cutting out a round piece. However, if we assume that the plasma forms only an outer shell around the center of the ball, the heating will be greatest at the rim of the circle (see Fig. 5.5 for an example). Now the stress created by the heating is very much concentrated at the outer edge of the circle and this makes it more likely that a roundish piece of glass might break free.[4]

---

[4] One should note that it is rather difficult to cut a round piece of glass out of a pane, even with good glass cutters.

BL Creation and Existence in Modern Aircraft

When Brand worked on his compilation of BL observations, he could not have included any cases concerning BL in aircraft, since aircraft had only just been developed during WWI to a stage where they could be used to transport people and some light goods like letters. Only later did fully metal aircraft appear that could then also fly higher in the clouds where some electrification could occur. Nowadays, the cruising altitude of jet aircraft is 10–12 km, higher than most weather effects; only thunderstorm clouds can go higher and these tend to be avoided because of their strong and turbulent winds. Lightning, on the other hand, is not a threat to aircraft: their conductive structure shields the interior with its passengers and electronics like a Faraday cage. Nevertheless, a considerable number of cases have been reported in which BL objects have been seen inside aircraft. Remarkably, the accounts are quite similar: the BL occurs at the front of the aircraft or in the cockpit, and travels down the aisle in the middle of the aircraft towards the rear, where it disappears. This book starts with an account of just such an observation by one of my acquaintances; another is given in full in the appendix (Case 19). This report is important because the observer was a scientist, Dr. Jennison, who gave a very detailed description of the object and the circumstances. In his case there was a lightning strike before he saw the BL.

The aircraft is, of course, not a completely closed metal structure. There are windows and antenna feed throughs, so it can at best be considered a somewhat leaky Faraday cage. A more detailed description of how lightning interacts with aircraft is given in the chapter on lightning.

The reports an BL objects in aircraft can be considered to be very reliable. The objects were seen at a close distance and the description is often very detailed.

Summary

- The first book on BL was written by Brand in 1923. For a long time its influence on BL research was quite small because it was in German.
- Brand used 215 reports which he considered reliable to extract the properties of BL observations in a statistical fashion, and most of his results are still valid today.
- The last book on BL, giving a complete overview of all aspects of the subject, was written by Mark Stenhoff and published in 1999.

- Compilations of BL observations are available from many European countries, the USA, and Japan.
- A. Keul analyzed these reports and concluded that they are based on a temporally and spatially stable core phenomenon.
- Brand's observation that BL is relatively more frequent in winter thunderstorms is supported by the newer data collections.
- The energy of most BL objects appears to be rather low. BL seen indoors may often store only a 100 J or even less.
- BL seen outdoors appears to be much more energetic, but here the energy estimates are difficult because the effects of normal, linear lightning, are often mixed with the effects of the BL object.
- The passage of BL objects through dielectric objects like glass windows is a well-documented feature that has been reported consistently over more than 100 years.
- There are a considerable number of BL reports from inside all-metal airplanes; they are remarkably similar with respect to the behavior of the luminous objects.

6

Electrical Discharges, Coronas, and Streamers

We have seen that the creation of ball lightning objects is closely related to normal lightning and especially to cloud-to-ground lightning. In order to understand the conditions that are created by cloud-to-ground lightning, we must take a closer look at electrical discharges in air: the corona, the streamers, and the leaders. St. Elmos fire, mentioned above, is just such a form of corona discharge.

Normal air is a good insulator. To make it electrically conductive, several things are required. To start with, we need to have free electrons. Electrons are very small and lightweight and can thus be accelerated easily, but they are normally tightly bound in atoms or molecules. So where do free electrons in the atmosphere come from? It turns out that in the atmosphere at sea level there are two factors which lead to free electrons: cosmic radiation and radioactive substances in and from the Earth. The atmosphere is constantly bombarded with energetic particles from space, mostly protons, but also gamma rays. These primary particles produce so called air showers, in which their energy is dissipated into a cascade of other particles, and some of them manage to penetrate our air shield down to the surface of the Earth. Additionally, the radioactive decay of uranium and thorium atoms in the Earth's crust creates Radon, a radioactive noble gas, which escapes from the rocks and reaches the surface. Its decay also contributes to the creation of free electrons in the air. In total, there are about ten free electrons per cubic centimeter of air at normal pressure. Since electrons are very light compared to atoms or ions, they can move much quickly in electric fields than positive or negative ions. In an electric field these electrons are accelerated, but at normal pressure, they tend to collide very often with neutral air molecules. Such collisions are often

© Springer Nature Switzerland AG 2019
H. Boerner, *Ball Lightning*, https://doi.org/10.1007/978-3-030-20783-0_6

elastic, meaning that the electron just bounces off a molecule, kicking it a bit and changing its direction, but not exciting it. Sometimes, if the electron is energetic enough, the molecule will be excited by the impact, and if the energy of the electrons is even higher, the impact will free another electron from the molecule, creating a new electron-ion pair. The energy which the electron can obtain in an electric field is determined by the field strength and the length of the path that the electron can move along without losing energy in a collision. The path between collisions is of course variable, but one can work with an average value: the mean free path of an electron in air of a certain density. If the pressure is lower, there will be fewer collisions and the mean free path will be longer, whereupon the electrons will be able to gain more energy from the accelerating electric field. When the field reaches a value where many electrons gain enough energy between collisions to ionize molecules in the next collision, the number of electrons will multiply and an electron avalanche is triggered. Such an avalanche is the prerequisite for generating an electric discharge in air. In air at normal pressure, a field of about 3 million volts or 3 MV/m is needed to achieve this condition. This is a very high field indeed, and we may wonder where such high fields could occur? In a plane geometry, like a flat countryside or a quiet water surface, the electric field created by a thunderstorm above (we will later see how a thunderstorm manages to be such a good generator of charge) is very homogeneous, but the introduction of a sharp conductive point will produce a concentration of the field at the place where the radius of curvature is smallest (see Fig. 6.1).

The reason is that the electrons can move freely in a conductor, and an electric field line which does not end perpendicularly on the conductor's surface will exert a horizontal force on the electrons, moving them until the horizontal force becomes zero and the field line has become perpendicular. This leads to a concentration of charges and field lines at the tip of a pointed

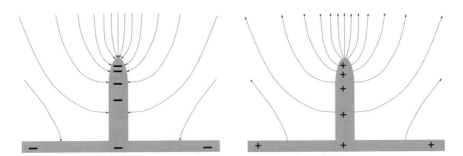

Fig. 6.1 Electric field enhancement at sharp conductive points. Drawing by the author

conductor and to an enhancement of the field in the vicinity of the tip. Figure 6.1 gives some indication, but it is not entirely correct: it was not possible to fit enough plus or minus signs into the tip, so you will have to imagine the concentration of charges there. The electric field close to a round tip is enhanced by a factor of 1/r, where r is the radius of curvature of the tip. With a tiny tip having a small radius of curvature, this ratio becomes very large, so one obviously gets high fields, and these then can form an electron avalanche and trigger a discharge.

If the tip is positively charged, electrons are created at the tip of the avalanche, and then they move towards the electrode, leaving a cloud of positively charged molecules in their wake (Fig. 6.2). These positively charged molecules, being much heavier and larger than electrons, are less mobile so they stay behind, creating what is called a space charge. The positive space charge forms an extension of the metal conductor, so the region where the field is high enough for ionization is extended forward, away from the tip, facilitating the growth of the ionized region.

In the case of a negatively charged tip, the situation is quite different. The electrons are now pushed away from the tip into a region where the field is lower, so they can no longer ionize the air. They also form a negative space charge away from the tip, and a positive one close to it, which leads to a concentration of the field close to the tip. Growth of a negative discharge is therefore more difficult than growth of a positive discharge. The free electrons at the outer boundary face another threat: they have little energy, so they can be caught by oxygen molecules to form negatively charged ions. Once again, these negative ions move slowly, so a considerable space charge is built up and in the case of a negatively charged tip, avalanche creation is stopped. When this negative space charge diffuses away, the field becomes strong enough to create new avalanches and the hole process repeats itself. This leads to very

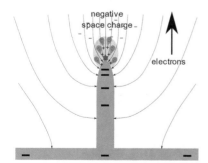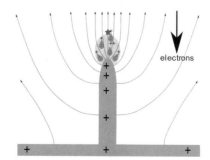

Fig. 6.2 Negative and positive coronas. The region of avalanche formation is shown in orange, and the region of the highest field with a red star. Drawing by the author

regular pulses called Trichel pulses in the case of a negative tip. Curiously enough, they were studied in connection with ball lightning in the 1960s (Pierce et al. 1960) because it was thought that they might be a source of electromagnetic waves. Sadly, it turned out that the highest frequency that can be obtained in this process is about 3 MHz, too low for the specific mechanism envisaged. We see that there are two processes involved: ionization by electrons and electron capture, and it is the competition between the two that determines whether the avalanche grows or not. The important factor is of course the strength of the electric field. If it is also strong enough in the regions further away from the tip where the field is enhanced, the discharge

Fig. 6.3 Corona discharge at a high voltage equipment at the National Bureau of Standards. Public domain

can propagate and form a tiny filament of conductive plasma which grows into the region of lower electric field. The discharge at lower fields is called corona discharge and at higher fields streamer discharge; these filaments are called streamers. Figure 6.3 shows an example in which corona discharge developed at several points along a high voltage transformer; the individual streamers are not visible because of the long exposure, but the shape of the discharges is clearly brush-like. The growth of streamers is away from the metal conductor, which supplies the current. Such discharges cannot be detached from the conductor and they do not have a clear spherical shape. For power lines, these discharges are a loss factor which needs to be avoided, so sharp edges with a small radius of curvature are not used. St. Elmo's fire is the same type of discharge, produced by the electric field of thunderstorms. Here the metal conductor is replaced by plants, trees, and sometimes animals and even humans.

The streamer tip now plays the role of the metal tip at which the streamer started out: it leads to a local field enhancement which facilitates the growth of the streamer. The ambient electric field for the self-propagation of streamers is about 4.5–7 kilovolts per meter (kV/m) for positive streamers but 10–15 kV/m for negative streamers. The streamers emit only a faint luminosity due to exited nitrogen molecules; in air the color is bluish. These streamers are not stable, coming and going quickly and producing a faint hissing or crackling sound. If the field is strong enough, they may extend further and further into the air. Several streamers now coalesce and form a luminous stem where the current is high enough to heat the air and create extra ionization. This stem is called the leader and because of its high conductivity, it helps the discharge to bridge the gap. In laboratory experiments, they are thus able to reach the opposite electrode. If that happens, they create a short-circuit between the electrodes because of their conductivity and the current jumps up, heating the air and generating an intense light emission: we say that a spark has jumped between the electrodes. The energy for the growth of the discharge is derived from the electric field: because of its conductivity, the leader is at the potential of the place it started from, and around the leader the field is now lower than it was before its growth. The discharge has thus converted the energy stored in the field into ionization and excitation of the molecules and heating of the air.

Streamers constitute a fundamental mechanism of discharges and have been quite well investigated and understood. The difference between the growth of positive and negative streamers and leaders has considerable consequences for lightning; we will come back to this point in the next chapter on lightning physics. The difference between positive and negative discharges also has profound consequences for BL production; we will discuss that in the chapter on BL experiments.

We have seen now that electrical discharges do not tend to produce luminous balls, but instead generate long filaments. Only when the corona or streamers are launched from an already spherically shaped conductor can we expect a ball-like appearance. Another difference is that ball lightning is described as a mobile object, whereas corona discharge is tied to sharp points like the top of a ship's mast, for instance. Ball lightning and corona discharges may appear under similar circumstances and they may even be related, but they are definitely different physical objects.

Finally, I would like to draw the reader's attention to a peculiar characteristic of BL surfaces: a number of observers describe the surface as being 'active' or being made up of 'worm-like discharges' (Case 12). If these luminescent fibers were streamers, the electric field creating them would have to be tangential to the object's surface.

Summary

- Corona discharges develop at points of high electric field which appear at points of conductors where the radius of curvature is small.
- Initially, coronas look like a bluish bundle of weak flame-like discharges.
- With higher electric field, the discharge develops into a streamer: the tip of the discharge channel provides a strong enough field to enable a self-propagating discharge to grow into regions of lower field strength.
- If the current increases further, the gas in the channel becomes heated and ionized. This channel emits more light and is highly conductive; it is called a leader.
- Corona is fixed to an object which has some electrical conductivity and can supply the current required for the discharge. Corona never develops in the open air or detaches itself from some object to move away into the open air.
- The discharge is different for positive and negative tips: for positive points, the electrons in the air are pulled into regions of higher field strength where they participate in more ionizing collision with air molecules. For negative points, the electrons are pushed into regions of lower field strength, where they cannot ionize more molecules and are instead attached to oxygen molecules, which therefore largely immobilizes them.
- Due to this different behavior, negative streamers need higher field strength to be able to propagate than positive streamers.

7

Thunderstorms and Lightning

We have seen that BL observations exhibit a clear correlation between thunderstorms, lightning, and BL, so we need to take a closer look at how lightning works in order to be able to understand BL. Since lightning research is a very broad and active field of research, we will have to restrict the discussion to the areas which are most likely to be relevant to BL. For a more detailed discussion of the many aspects of lightning physics, there is an excellent review of the current state of the art (Dwyer and Uman, 2014), and the books by Cooray (2015) and the "lightning bible" of Rakov and Uman (2003) cover all the scientific aspects of lightning and lightning protection.

How Thunderstorms Power the Electric Machine

The action starts on a nice summer's morning with the Sun warming the wet ground. The warm, moist air rises upward, because it is less dense than the colder air around. As it moves upwards, the moist air cools down bit by bit. At a certain height, the temperature is so low that the water vapor condenses around small dust particles and forms tiny water droplets: a cloud forms in the air. The condensed water releases the energy that was required to evaporate it, heating the air around it a bit more and increasing its buoyancy. With this increased uplift, the parcel of warm air shoots up higher and higher. When the supply of warm, moist air from the heated ground is sufficient, the cloud can reach a height at which the temperature falls below freezing. The water droplets become supercooled, which means that they are still liquid even though their temperature is below freezing. Then some of them collide and form

© Springer Nature Switzerland AG 2019
H. Boerner, *Ball Lightning*, https://doi.org/10.1007/978-3-030-20783-0_7

small ice crystals, creating what is called Graupel. Graupel is basically made up of ice crystals frozen onto a snowflake. These Graupel particles collide with supercooled water droplets and other Graupel particles in the turbulent updraft of the burgeoning storm cloud. During these collisions, the particles become charged. The smaller particles carry the positive charge, whereas the larger ones become negatively charged. These opposite charges attract each other but are separated by the updraft that carries the lighter, smaller particles up faster than the heavy ones that stay behind. In this way, the updraft separates the opposite charges, depositing energy in the electric field of the developing thundercloud. The Graupel particles continue to grow until the updraft can no longer lift them, at which point they start to fall towards the Earth. Some of them melt in the warmer air, but the larger ones may reach the ground as ice particles: hailstones are falling. Depending on the strength of the updraft and the time they had for growth, the hailstones can become very large. The largest ones on record were more than 10 cm in diameter.

The smaller particles are carried high up until they reach the upper limit of the troposphere, and there they hit the next layer of the atmosphere, which is called the stratosphere. Here the temperature profile changes: it now gets warmer as one goes further up. This blocks the updraft and the cloud can only grow in the horizontal direction, forming the typical structure of mature storm clouds: the anvil. The anvil contains small ice particles carrying positive charge. Figure 7.1 shows a photo of a mature tropical thunderstorm cloud. Its top is considerably higher than the cruising altitude of the airplane from which the photo was taken.

A schematic view of such a storm cloud is shown in Fig. 7.2. The main concentration of negative charge is at a level in the cloud where the temperature is between −10 and −15 °C. The main positive charge is high up in the anvil. The updraft generated by sunshine powers this huge engine, but for isolated thunderstorms, it cannot be sustained for long. In the evening, when the supply of warm, moist air becomes less abundant, the thunderstorm becomes weaker and in the end it dissolves. Such a single, isolated thunderstorm cell is typical of many summer thunderstorms in temperate regions. But there are other ways a thunderstorm can be started: many of the more violent thunderstorms are produced by the inrush of cold air into a mass of humid air, lifting it up. Such a situation not only produces one thunderstorm cell, but a whole line of cells, so-called supercells, that can extend over hundreds or even thousands of kilometers.

The typical charge structure of a thunderstorm consists of this charge dipole: the positive charge on top and the negative charge below. In many cases, a smaller pocket of positive charge exists below the negative charge res-

Fig. 7.1 Tropical thunderstorm cloud. Photo taken from a jet plane at cruising altitude. Photo by the author

ervoir. Of course, large thunderstorms do have a more complicated charge distribution, but the thunderstorm dipole or tripole helps to explain most of the features of lightning we are interested in. In winter, thunderstorm clouds are lower and the wind in the upper regions is often stronger than below, producing a wind shear which exposes the upper, positive charge to the ground. Under certain, rarer circumstances, an inverted charge structure can also appear (Fig. 7.3).

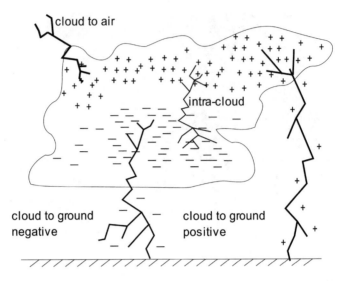

Fig. 7.2 Thunderstorm cloud with different lightning types. Drawing by the author

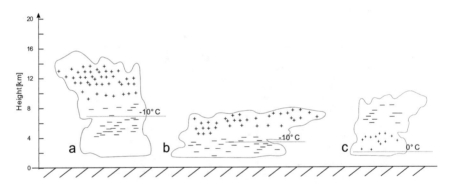

Fig. 7.3 Thunderstorm clouds: **(a)** summer thunderstorm, **(b)** winter thunderstorm with wind shear, **(c)** winter thunderstorm with inverted charge structure. Drawing by the author

The Members of the Lightning Family

From the dipole charge structure of thunderclouds, most lightning will obviously occur inside the clouds, between the main positive and negative charge centers. This intra-cloud lightning accounts for about 90% of all lightning discharges. When there are several thunderstorm clouds near to each other, lightning between clouds, or inter-cloud lightning, is possible. Friends of mine once observed a brilliant display of lightning in the western Amazon region, when almost constant lightning jumped between several huge pillars of thunderstorm clouds. A rather rare type of lightning is from the cloud to the surrounding air, known as cloud-air lightning.

Most of the cloud-ground lightning comes from the lower, negative charge center. This −CG lightning accounts for roughly 90% of all lightning to the Earth. Only 10% of the lightning to ground is from the positive charge on top of the thunderstorm cloud. This positive charge is normally well shielded from the ground by the intervening negative charge. Only when the anvil spreads out far beyond the clouds underneath can +CG lightning travel all the way from the top of the thunderstorm cloud to the Earth.

Another classification is due to Berger, a lightning researcher from Switzerland (see Fig. 7.4).

He worked with an instrumented tower in the Alps and therefore realized that lightning can also start from high buildings on the ground, hence the two categories for upward-going lightning.

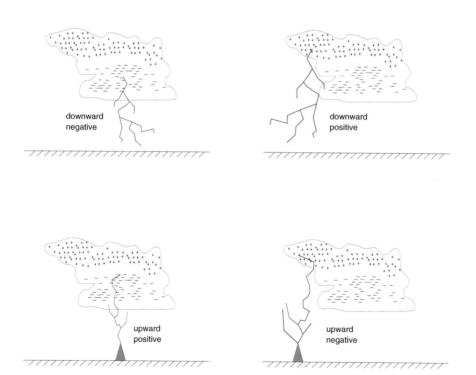

Fig. 7.4 The four lightning types according to Berger. Drawing by the author

CV of a Typical–CG Lightning Stroke

A typical negative cloud-ground lightning stroke starts deep inside the storm cloud with an initial breakdown of the air. The curious thing about this is that nobody really knows how it works. The problem with the beginning of the breakdown is that the electric fields measured by probes in storm clouds are not strong enough for an electrical discharge. At ground level, about 3 MV/m are required, at a height of 5 km, about half of this value, and even less in the environment of a cloud, but still fields that are enough to trigger a discharge have never been measured. A field enhancement mechanism must be operating in order to trigger lightning. It is of course difficult to measure the electric field in such a hostile environment as a thunderstorm cloud, so small pockets of higher fields may have been missed. In these pockets, collisions of particles may produce fields that can start a local discharge, that then grows into a streamer. This is the classical field enhancement explanation. A second one is the so-called relativistic breakdown theory. To explain how this works, we must take a look at electrons moving in air under the influence of an electric field. The field tries to accelerate the electrons, but they continuously bump into the molecules of the air, losing energy in each such collision. The effect is like another force, counteracting the electric field, basically dragging the electrons in such a way that they only drift with a constant velocity. This friction force is not constant, but increases with the speed of the electrons, making their acceleration an uphill battle. Take a look at Fig. 7.5 to see the form of the relationship between the electron speed and the friction force.

The interesting thing is now that this force does not always increase. There is a maximum. Beyond that peak, faster electrons experience a smaller force, so now their acceleration is easier! When the electric field is strong enough and the space filled with the field is large enough, the electrons experience a runaway acceleration until they reach relativistic speeds, that is speeds that come close to the speed of light. Then several interesting things can happen. Basically, the electrons can be scattered by nuclei of air molecules, generating very energetic gamma-rays which can in turn produce electron–positron pairs in other scattering events. But even before they reach these high energies, they create secondary electrons by knocking them away from neutral atoms. The field of the thunderstorm cloud acts here like a huge accelerator, creating an avalanche of electrons that facilitate the start of a discharge. The only question is how to get electrons with speeds in the runaway region. Here this theory takes another exciting twist: in fact, electrons of the required speed are available high up in the atmosphere. They are created in so-called air showers by

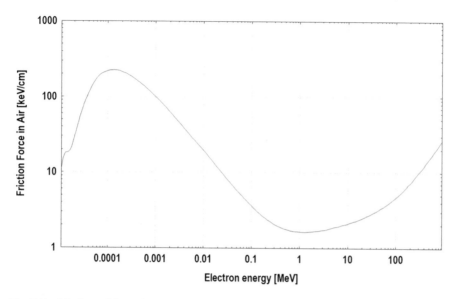

Fig. 7.5 Friction of free electrons in air. Diagram by the author

cosmic rays. The Earth is constantly being bombarded by protons and heavier nuclei, as well as energetic gamma rays. In the upper atmosphere they collide with the nuclei of air molecules and create a cascade of other particles. In these cascades there are a number of electrons or positrons of sufficient energy for relativistic runaway avalanches (Fig. 7.6).

To sum up, the whole process of triggering a lightning stroke involves the following steps:

- Charge separation in the thunderstorm cloud creates an electric field of about 1/10 of the breakthrough strength in a sufficiently large volume, say at least several hundred meters or even several kilometers. Then there is enough space to accelerate electrons or positrons.
- A cosmic ray particle hits the nucleus of an air molecule, starting a cascade of secondary particles, if it is a proton or a heavier nucleus. Alternatively, a high-energy gamma ray from space starts an air shower. In both cases, high energy electrons or positrons are created.
- The electrons or positrons are accelerated in the electric field of the cloud, creating an avalanche of secondary electrons.
- The charge of the avalanche is sufficient to create a breakthrough, producing a streamer, the starting point of the lightning discharge.

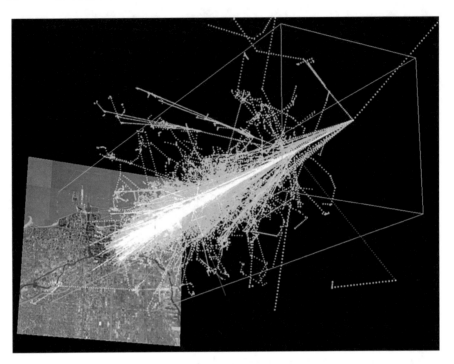

Fig. 7.6 Simulation of an air shower. The primary particle enters from the right. From: Original uploader was Dinoj at en.wikipedia (https://commons.wikimedia.org/wiki/File:Protonshower.jpg), "Protonshower", CC BY 2.5

The streamer is now self-propagating, since it creates very high fields at its tips. But there is a difference between this streamer and the streamers we discussed above: here in "thin air", there is no conductor supplying current to the streamer. The streamer must supply its own current, and that is only possible if it is bipolar: on one end a positive streamer tip is growing, producing and collecting free electrons. These are then channeled along the well conducting filament of the streamer to the negative tip growing in the opposite direction, and here the electrons are emitted into the air. If the electric field is strong enough and extends over a sufficient region of space, the streamer will continue to grow, taking more current and growing into a leader, where the filament is heated by the current and becomes even more conductive. You may ask yourself if this scenario is realistic and if there is any evidence for it. The answer is affirmative: since 1994 gamma-ray emission has been observed from thunderstorms. An experiment on board the Compton Gamma-Ray Observatory, a NASA satellite, detected gamma-ray bursts, not from space as expected, but from Earth. These high energy bursts were linked to lightning in thunderstorms beneath the satellite. Meanwhile, it has become clear that

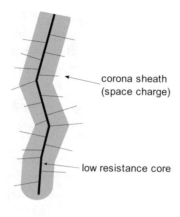

corona sheath
(space charge)

low resistance core

Fig. 7.7 Lightning channel with corona sheath. Drawing by the author

these gamma rays are really emitted from within thunderstorm clouds and are due to some relativistic runaway mechanism, but the exact details are still being investigated (Dwyer and Uman, 2014) (Fig. 7.7).

In cloud-ground lightning, this leader leaves the cloud and propagates towards Earth. As everyone knows, such a leader does not usually have a single discharge channel. Figure 7.8 shows a number of branches growing out of a cloud, as though probing the space between cloud and ground and testing where it would be easiest to move forward. This stage of lightning looks like the branches of an uprooted tree, or like the roots of a tree. In negative cloud-ground flashes, these leaders do not advance smoothly, but propagate in a stop-and-go fashion, whence they are called stepped leaders. Each step advances the leader by 10–100 m, and the interval between steps is about 10–100 μs. This curious behavior is due to the fact that the electrons at the tip of the leader have to move into the lower field region ahead, where they cannot start an avalanche mechanism. Therefore, a considerable charge accumulates at the tip, eventually creating a high enough field to create a new bidirectional streamer ahead of the existing leader channel. The connection to the leader looks like a jump ahead of the leader itself. The conditions at the tip of the stepped leader are quite extreme. As recently as 2005 it was shown that these leader tips produce X-rays, indicating a relativistic runaway mechanism here as well (Dwyer et al., 2005). The growth of stepped leaders is best seen in slow-motion videos.[1] The core of these leaders is a few centimeters in diameter, and because of the rather strong current flowing, it has a temperature

---

[1] You can find a number of such videos on, e.g., YouTube, by searching for ultra-slow-motion lightning.

between 10,000 and 20,000 K. The high conductivity of this core means that the electric potential of the thunderstorm cloud, which can be several million volts, is guided down towards Earth. The high voltage produces a corona around the leader's core and branching of streamers and leaders. This region around the channel of the leader is called the corona sheath; it stores a large amount of charge, rather like a very long cylindrical capacitor. Its diameter is between a few meters and a few tens of meters, depending on the charge stored in the sheath.

When these leaders approach the ground, the electric field increases and at first some corona discharge starts at sharp pointed objects. When the leaders come closer than about a few hundred meters, streamers appear which can grow into leaders from the ground, approaching the descending stepped leader. Usually, there are several upward growing leaders, but in the end, only one of them comes so close to the stepped leader that, with a final jump, the distance between them can be closed and cloud and ground become electrically connected. Then a current pulse moves upwards at one third to one half the speed of light and the charge stored in the corona sheath is drained to Earth. This is what we really see as lightning, since now the current in the leader channel reaches thousands of amps, and as a consequence the light emission goes up dramatically, the leader channel explodes, and a shock wave is emitted which we hear as the thunder clap. This discharge is called the return stroke. In negative cloud-ground lightning, the discharge stops when the charge in the leader channels is gone. Since the charge distribution on the thunderstorm cloud is now changed, other pockets of charge may become connected in turn to the remains of these leaders, starting new leaders towards Earth. These so-called dart leaders take advantage of the remnants of the return stroke, which contains ions that have not been able to recombine and heated air around the old lightning channel. They are therefore able to move faster than the stepped leaders and most of the time they do not display a stepping behavior. They more or less follow the trace of the previous return stroke and do not usually have branches like the stepped leader. The sequence dart leader—return stroke can repeat itself several times, until all the charge pockets in the cloud which can be accessed by the lightning have been discharged. Up to 25 individual strokes have been recorded in a single flash of negative CG lightning. The individual return strokes follow each other so fast that we see this as a flickering of the main lightning channel.

The attachment process, in which the upward leaders reach for the downward leader, is of considerable importance both for lightning protection and for BL research, so we shall take a closer look. In Fig. 7.8, two upward leaders

Fig. 7.8 Upward leaders from two high buildings. From video on YouTube (https://youtu.be/vbQtqsolWjs). With kind permission of Marcelo Saba—CCST/INPE

are issued from the lightning rods of two moderately high buildings.[2] The leader from the right-hand building wins the race and connects to the stepped leader above, starting the return stroke. The left upward leader just fizzles out. This spectacular image shows that lightning does not really "hit" an object on the ground, there is always a leader growing upwards to meet the downward leader. The resulting return stroke is a wave front moving upwards. Of course, the impression is that the lightning comes down to strike something on the ground with a physical force, so in ancient times it was supposed that a material object called a "thunderbolt" was thrown from heaven to Earth to wreak destruction there.

The length of the upward leader depends on the charge in the stepped leader which creates the electric field at ground level and the height of the object issuing the upward leader. A larger charge will create a stronger electric field and a tall object will be closer to the stepped leader. Lightning rods, in fact, are meant to start the upward leaders so that the discharge is guided to a point where it can do no harm. Unfortunately, the path of downward moving and upward moving leaders is quite erratic due to unpredictable details of the discharge process, so it is not always guaranteed that the leader from the lightning rod will win the race. For this reason, a number of devices have been marketed which claim to be able to start the upward leader early, so as to

---

[2] The image is one frame from: https://youtu.be/vbQtqsolWjs. See also (Saba et al., 2017).

ensure a connection to this lightning terminal. These are the so-called early streamer emission terminals (ESE terminals). Unfortunately, their effectiveness has never been demonstrated, so a good old blunt lightning rod is still the best and cheapest lightning terminal you can get (Cooray, 2015). The distance between the tip of the object or lightning rod that started the upward connecting leader and the tip of the stepped leader at the time of the final connecting jump is called "striking distance". As mentioned above, stronger charge leads to longer striking distances and larger maximum current in the return stroke (which discharges the corona sheath), so there is a correlation between maximum current and striking distance. Stronger lightning is more easily caught by lightning rods, at least when we are talking about negative CG lightning. The situation is rather different for positive CG lightning, which we will investigate now.

Positive CG Lightning

Positive cloud-ground lightning is much rarer than its negative cousin. The negative charge layer underneath the upper positive charge layer in thunderstorms usually shields it very well against the ground, so lightning from the positive layer ends up as intra-cloud lightning. When the upper positive charge is more exposed to the ground, positive CG lightning can occur. This is the case at the end of thunderstorms, when the lower parts of the clouds have moved on or have partially dissipated, but the far-flung anvil with its positive charge is exposed. Then quite spectacular lightning from the anvil can propagate down to the ground from a height of about 10 km. On average, only about 10% of all CG lightning is positive. The leader of a +CG lightning stroke does not step: it moves down in a more or less uniform fashion. Moreover, its tip does not create X-rays as the negative counterparts. Positive lightning flashes are almost exclusively single-stroke events. There is no dart leader—return stroke sequence as with −CG lightning, but in contrast, +CG lightning very often involves a long-lasting current (known as the continuing current). Obviously, the positive lightning discharges its charge reservoir completely in one go. The charge channeled to Earth can be much higher than with −CG lightning (Rakov, 2003), and in addition, +CG lightning tends to involve much higher maximum currents. Up to about 400,000 A have been measured, about twice the maximum recorded for −CG lightning. In particular, the long-lasting current at the end of the +CG stroke can heat substances to above burning temperature, so it is no wonder that +CG lightning can do quite spectacular damage, much worse than −CG lightning. Trees hit by posi-

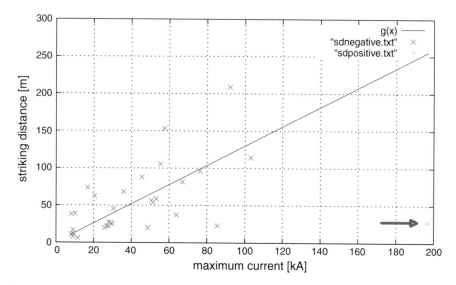

Fig. 7.9 Striking distance as a function of the maximum lightning current. Drawing by the author, data are from Wang et al. (2016)

tive lightning often completely explode because the water in the tree is vaporized, whereas negative lightning may only strip some bark off the tree it hits (Heidler et al., 2004).

The attachment process also involves an upward growing leader, but here the polarities of the leaders are reversed: the downward positive leader is able to move smoothly, but the upward leader is now a negative leader; here the electrons at the leader tip move into a region with lower field and—as explained above—in order to start and propagate, these negative streamers and leaders need an electric field which is 2–3 times higher than the electric field for a positive leader. Therefore, the striking distance is lower for positive CG lightning. A study of the striking distances for negative and positive CG lightning (Wang et al., 2016) shows this clearly: the only positive stroke had a very high current but the striking distance was about a factor of 10 lower than for negative strokes (see Fig. 7.9, where the point indicated by the red arrow is from the positive stroke). The violet line is a fit to the distance values of the negative strokes.

This has two consequences: in order to catch the dangerous positive CG strokes reliably, lightning rods must be spaced more densely than for negative CG strokes and the energy density of the electric field at the ground can be almost 10 times higher for +CG than for −CG, because the energy density goes as the square of the field strength.

As mentioned above, positive CG lightning tends to occur towards the end of thunderstorms, but there are other situations where it can occur more frequently than in normal thunderstorms.

It has been known for a long time that winter thunderstorms tend to have a higher percentage of +CG strokes than summer thunderstorms.

In winter thunderstorms, the cloud height tends to be lower than in summer, because of the weaker convection, but there is often a strong wind blowing at the height of the thunderstorm clouds and not at ground level. This wind shear tends to displace the upper, positively charged part of the cloud ahead of the lower, negatively charged parts. The positive charge is then only weakly screened from the ground or not screened at all, and positive CG strokes become possible. Sometimes this wind shear leads to bipolar lightning patterns: downwind one finds the positive CG strokes and upwind the negative ones (see Fig. 7.3). This explanation cannot be used to explain all positive lightning in winter; in Neuruppin (Case 1), the strong positive lightning was more likely produced by a thunderstorm cloud with an inverted structure, where the positive charge was below the negative.

Positive lightning of considerable strength can even occur in quite isolated cases. This so-called rogue lightning often occurs in winter, but also in summer. Insurance companies used to regard lightning damage claims from isolated strokes in winter with considerable skepticism, but when lightning location systems became available, it turned out that such isolated strokes do indeed occur (Holle et al., 1997; Idone et al., 1984) and are often positive CG strokes.

Other situations are also likely to lead to a higher percentage of +CG lightning than normal. When air containing smoke from bush fires is incorporated into a thunderstorm cloud, the percentage of positive lightning has been seen to go up (Rakov, 2003). This is not so surprising, because the extremely fine smoke particles act as condensation nuclei and can alter the charge generation structure within the thunderstorm cloud. Hurricanes and tornadoes have also been linked to the generation of positive lightning. A recent study of lightning in very strong North Atlantic hurricanes (Thomas et al. 2010) showed that phases of intensification and especially calming can be accompanied by lightning with a high ratio of positive strokes.

Other Types of Lightning

Most CG lightning has always been started by leaders coming down from the clouds, but this is not always the case today. Figure 7.10 shows probably the first photo of lightning going upwards from high buildings, here from the

THE EIFFEL TOWER AS A COLOSSAL LIGHTNING CONDUCTOR.
Photograph taken June 3, 1902, at 9.20 p.m., by M. G. Loppé. Published in the *Bulletin de la Société Astronomique de France* (May, 1905).   [*Page 82.*

Fig. 7.10 Upward-moving lightning from the Eiffel tower. Photo from 1902, public domain

Eiffel tower, which was quite new at the time the image was taken and was in fact the highest building in the world.

Upward-moving lightning must have been quite rare before high buildings became common, but it is now routinely observed. Many more images can be found on the internet. The fact that these buildings both start their own lightning or are effective in collecting down-coming lightning makes them good places for lightning research. A wooden tower on top of the Monte San Salvatore near Lugano in Switzerland was used by the lightning researcher Berger to measure the current of lightning strokes (Berger, 1973), but now towers in Germany (Hoher Peißenberg), Austria (Gaisberg), and several other places around the world are used for lightning experiments. These towers have instruments to measure the current shape of the lightning, while high-speed video recordings are usually used to capture the development of the discharge. Since the place and the strength of the lightning is known, this information can be used to calibrate the data of lightning location systems. For BL investigations, there is one difference between CG lightning from high buildings and CG to the ground: the towers are well grounded and the upward-going leaders are basically also at the ground potential because of their good conductivity. Therefore, the electric field at or around the tower is never as high as in the case of a leader approaching the ground.

When no high building or tower is available, but you would nevertheless like to study lightning, you can use a different method: fire a rocket up into a thunderstorm cloud and trigger lightning! The rocket needs to be connected to the ground by a conductive wire, so this basically simulates the effects of a high building. Once again, since one knows where lightning will strike, instruments for recording currents can be used. Also other effects can be studied: the emission of X-rays by stepped leaders was first seen by equipment placed near the launch site of just such a rocket (Dwyer et al., 2005). The situation with respect to the maximum electric field around the strike point is like the upward-going lightning from towers: the field does not reach as extreme values as for normal CG strokes. Only when the rocket uses a trailing wire not connected to the ground does the lightning triggered begin to look more like normal CG lightning.

In both cases, i.e., upward lightning from tall buildings and lightning triggered with grounded wires, the discharge is different from downward lightning. The upward growing leader is supplied with charge from the well-conducting ground, but when it reaches the space charge in the cloud, no return stroke is initiated. The cloud is simply not conductive enough for such a discharge. It is possible, on the other hand, that new discharges will be initiated by subsequent dart leaders coming down. In these cases, return strokes are observed.

The triggering by a rocket with a trailing wire is comparable to lightning triggered by aircraft, see Fig. 7.11. Here both positive and negative leaders are started from the conductive airplane in opposite directions; the direction of growth of one of the leaders can be seen by the branching which is directed away from the aircraft. This is the most common case where lightning "strikes" an aircraft; approaching leaders appear to be the rarer case (Rakov and Uman, 2003).

The last type of lightning we will look at is volcanic lightning. It has been known since antiquity that some volcanic eruptions can produce lightning in their eruption columns. Usually, soon after a burst from the eruption vent, lightning discharges from within the dense, cooling ash cloud, sometimes also striking the area around the crater (Fig. 7.12). Obviously, in these hot ash clouds an effective charge separation mechanism must be working, but the details are not known. It is simply too difficult to study what is going on

Fig. 7.11 Aircraft triggering bipolar lightning. Drawing by the author, using: Cvdr (https://commons.wikimedia.org/wiki/File:Airbus_A320_with_satellite_antenna_radome.svg), "Airbus A320 with satellite antenna radome", CC BY 3.0

Fig. 7.12 Volcanic lightning near Rinjani volcano, Indonesia, in 1994. From: Oliver Spalt (https://commons.wikimedia.org/wiki/File:Rinjani_1994.jpg), "Rinjani 1994", CC BY 2.0

within such an eruption cloud! Nevertheless, it turns out that the net charge emitted from the volcano is positive (Thomas et al., 2007), so positive charge is dominant in the ash plume extending away from the volcano. Once again, the details of charging in the plume are being studied, since many factors may be involved. For a recent paper see Harrison et al. (2010). If we are concerned with BL studies, the essential fact is that the ash plume extending away from the eruption vent tends to be positively charged. This fact will become important when we look at one of the more interesting BL reports later.

Luminous Events in the Upper Atmosphere

There are several types of luminous events high up in the atmosphere, and some of them produce quite spectacular displays. Figure 7.13 shows an overview.

We have already considered sprites in the context of natural phenomena which were long ignored by science. With low-light level cameras becoming more widespread, acquisition of sprite images is now not only a research activ-

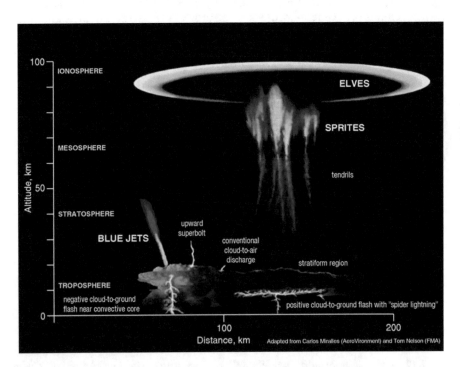

Fig. 7.13 Luminous event in the upper atmosphere. Public domain

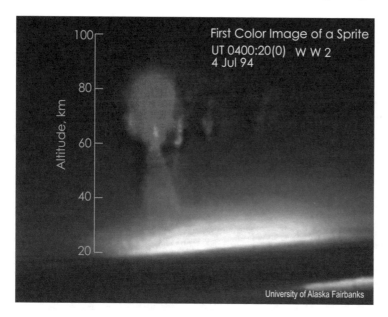

Fig. 7.14 Big red sprite. Public domain

ity[3]; quite a few hobbyists are actively hunting for them as well. Sprites are red discharges occurring at altitudes of 50–90 km. It is now known that sprites are just something we have already encountered, but on a much smaller scale: they are streamers which achieve this enormous size of many tens of kilometers in the thin air of the upper atmosphere. They are triggered by the disturbance of the charge balance which happens when a lightning stroke transfers a very large amount of charge from high up in the thunderstorm cloud to the ground. The resulting potential difference between the top of the thunderstorm and the conductive ionosphere is what drives these beautiful discharges. The vast majority of lightning strokes that have been observed to trigger such sprites are strong positive CG strokes. Only a very few negative CG strokes have been recorded triggering sprites. For a detailed and comprehensive study of sprite triggering lightning, the reader is referred to the paper (Williams et al., 2012). Since strong positive lightning often originates from the top of the thunderstorm and since it often discharges a lot of charge in the continuing current, we should not be surprised that they are the main reason for sprites. For our purposes, it is interesting to note that observation of sprites is in turn a good indicator of strong positive lightning (Fig. 7.14).

---

[3] Search, for example, for Martin Popek.

Fig. 7.15 Gigantic jet. Public domain

Other spectacular luminous phenomena are blue jets and their cousin, the gigantic jets, like the one shown in the Fig. 7.15. These are huge discharges going upward from thunderstorm clouds, the gigantic jets reaching the altitude at which sprites usually occur, where they change their color from blue to the red of sprites. Blue jets are rarer than sprites and their origin is still not clear.

Lightning Location Systems

Lightning is not really trying to hide its presence: its light can be seen from far away, even from space, and thunder can be heard over many kilometers. In the early days of meteorology, thunderstorm days were really counted by the presence of thunder. Nevertheless, a better way of detecting and locating lightning at greater distances and with greater precision was needed for several reasons, especially in the case of cloud-to-ground lightning: severe weather forecasting needed information about the formation and movement of thunderstorm cells, insurance companies wanted to check the damage claims of customers, and utility companies wanted to get rapid information about where lightning had struck high voltage overhead power lines in order to be able to fix the damage quickly.

It has been known since the beginning of wireless transmission that the huge sparks of both intra-cloud and cloud-ground lightning emit electromagnetic waves. Cloud lightning tends to emit more in the high frequency range, with frequencies of 1–10 MHz, whereas CG strokes due to the strong return stroke current change emit like kilometer-long vertical antennas with a maximum at frequencies of 10–20 kHz. These waves with long wavelengths of 10–100 km can travel very great distances. The ground wave, propagating along the surface of the Earth, goes the shortest distance and arrives first, but there are reflections from the conductive ionosphere, which arrive later. Figure 7.16 shows what this looks like. Over land, the ground wave is attenuated after about 1000 km, but over oceans, where there is much less attenuation due to the high conductance of the sea water, it can travel much further. Furthermore, reflection on the surface of the ocean works better, so at least strong strokes can be detected over many thousands of kilometers. Later, we will encounter the reflection of electromagnetic waves by a plasma, as in the ionosphere, in a different context, since a reflecting plasma shell is needed for a certain class of BL models. In these models, the plasma reflector encases a microwave field.

In order to locate the foot point of CG discharges, two methods have been used. The first one, the magnetic direction finder, detects the direction from which the stroke signal arrives. It receives the magnetic field emitted by the return stroke with two perpendicular antennas. The ratio of the signals on these antennas gives the direction of the point where the signals were emitted. If two or more stations receive signals from the same stroke, the position of the stroke can be triangulated.

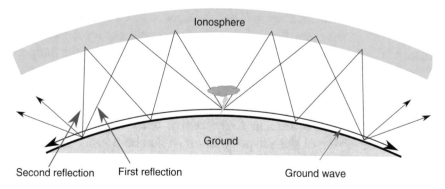

Fig. 7.16 Ionospheric reflections of the lightning impulse. Drawing by the author

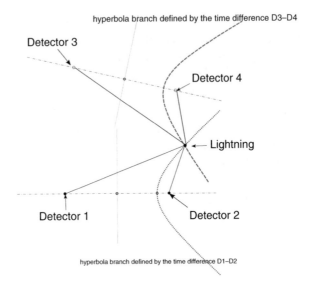

hyperbola branch defined by the time difference D3–D4

Detector 3

Detector 4

Lightning

Detector 1

Detector 2

hyperbola branch defined by the time difference D1–D2

Fig. 7.17 Lightning location using the time-of-arrival (TOA) difference. Drawing by the author

The second method works with the time of arrival of the stroke signals (Fig. 7.17). Since electromagnetic waves propagate at the speed of light, very precise timing and very good clock synchronization is required at the different receivers. The signals propagate 300 m in 1 μs, i.e., in one millionth of a second, and that's the minimum timing precision one must aim at. This accuracy was hard to achieve before the GPS became operational. Indeed, GPS is used not only to determine the position of the receiver, but also to transmit a standard time signal, which in most receivers gives a timing accuracy of about 0.1 μs. Lightning location works as follows. Let's assume we have two receivers recording the signal from a stroke occurring exactly halfway between them (Fig. 7.16). Then the receivers will get the same arrival time for the signals, since they travel the same distance to each of the receivers. Consequently, the source of the signals must lie on a straight line perpendicular to the midpoint of the line connecting the receivers. All points on this line would give the same time difference of zero. If the strike point is closer to one receiver, there will be a time difference, so the locus of possible points is no longer a straight line, but a hyperbola. If there are four receivers, one can get the exact point from the intersection of these hyperbolas for strokes both inside and outside the area surrounded by the receivers. Besides the exact timing, a receiver for such a network must be able to digitize the signals to extract the correct point for the time measurement, and it has to send the data to a server which combines the information from the many sensors that comprise the network. Electronics and digital networks only became widely available in the 1980s, and in this period, a lightning location system spanning the whole of the continental United States was set up. In the 1990s, many other networks went into operation, first in individual countries like Austria and Germany. Since 2000, there has been a partnership between several national European networks, called EUKLID, which currently consists of about 147 sensors in 27 countries. Many other networks are now operational over the whole world. All detect stroke position and time, and many also extract the strength of the stroke via the maximum current, and also the polarity of the stroke. At least one of them locates strokes over the whole of the Earth.[4] The location accuracy of such a time-of-arrival system depends mainly on the timing accuracy, and within a network of stations, it can be on the order of few hundred meters when several corrections are taken into account (Cummins and Murphy, 2009). The stroke polarity is determined by recording the electric field of the stroke, for example, with a short vertical antenna. The strength is extracted from the maximum signal, which can be correlated to the maximum current of the return stroke. Calibration of the location and the maximum current can be derived from strokes to instrumented towers, or using rocket-triggered lightning where place and strength are known.

Thanks to these lightning detection networks, many different kinds of information have been collected. To name but a few:

- Seasonal variations: lightning in summer is concentrated over land, because here the convection is stronger due to the warmer land mass, but lightning in winter occurs mainly over the ocean due to the warmer water driving convection (Pedeboy et al., 2017).
- Winter thunderstorms over land tend to have a higher percentage of positive lightning strokes than summer storms (Rakov, 2003).
- Strong negative and positive CG strokes are produced by different weather systems (Lyons et al., 1998).
- Strong negative CG strokes are more prevalent over the ocean than over land (Said et al., 2013).
- Exhaust fumes from shipping create more lightning along frequently used shipping lanes (Thornton et al., 2017).

In addition to the TOA systems, there are also satellite systems covering the whole Earth. These are used for global lightning incidence statistics. On a

---

[4] See http://wwlln.net/

local scale, receivers operating in the VHF band (50 MHz to several hundred MHz) record the emissions of lightning channels inside clouds and allow for a 3D reconstruction of intra-cloud lightning activity. These systems can record the complete spatio-temporal development of discharges. One particularly interesting system is LOFAR (LOw Frequency ARray). This is a novel radio interferometer, a radio telescope in the Netherlands which consists of a large array of simple and cheap antennas recording signals below 250 MHz. The signals are recorded individually and later processed in a large computer to extract whatever information is desired. During thunderstorms, it also receives signals from the lightning in thunderstorm clouds. Recently, these signals have been analyzed and the lightning channels could be reconstructed to an accuracy of about 1 m (Hare et al., 2018). Since LOFAR can also record the weak radio emission originating from air showers produced by cosmic rays, it may be possible to check the hypothesis that lightning is initiated by the ionization produced by cosmic rays.

For the investigation of BL, lightning location systems are invaluable. They can be used in two ways: when a BL is reported, one can quickly check whether any strokes were reported close to the position of the observer. If yes, distance, strength, and polarity can be obtained.

The second possibility is to identify special strokes that may have given rise to especially spectacular BL production, and ask the public whether anything unusual has been observed.

Armed with this knowledge of lightning physics, we can now proceed to the description of particularly spectacular BL observations that can give us a first clue about the circumstances in which these objects are created.

Summary

- Thunderstorms are driven by convection, a force moving moist air several kilometers up into the atmosphere.
- Positive and negative charges are separated by collisions between graupel particles.
- The smaller ice particles become positively charged during the collision and are hoisted to the top of the thunderstorm.[5] Larger, heavier particles obtain a negative charge.
- Negative charge accumulates below, forming a charge dipole in the cloud.
- Most lightning occurs within the cloud, between different charge pockets.

---

[5] Details of this charging process are still unclear.

- The actual origin of a lightning discharge is still not clear, because the electric fields inside thunderstorm clouds are a factor of 10 lower than the necessary breakdown fields.
- A popular theory proposes electrons created by cosmic rays as a lightning starter.
- Electrons are accelerated to almost the speed of light and create gamma rays and X-rays in their collisions.
- Cloud-ground lightning (CG) comes in two forms: negative and positive.
- Negative CG lightning (−CG) is by far the most common, constituting about 90% of all strokes.
- Positive CG lightning (+CG) often comes from the anvil, the topmost part of the thunderstorm cloud, and travels many kilometers before reaching earth.
- +CG lightning can be much stronger than −CG lightning in terms of the maximum current and continuing current, so it is often more destructive.
- The percentage of +CG lightning is higher in winter thunderstorms, but also thunderstorms that have ingested smoke from bush fires, and in certain particularly violent thunderstorms.
- Luminous discharges above storm clouds—sprites, elves, blue jets, and gigantic jets—have become a focus of research since about 1990.
- Lightning location systems are becoming an essential tool in linking BL observations to physical processes.

8

BL: Well Documented Cases of Copious Production

We have now collected almost all the information that is available on BL. Especially important is the curious fact—already noted by Brand—that BL is created more frequently in winter than one would expect from the very few lightning strikes in that season. This fact is still observed in other and more recent collections of BL observations. He also noticed that BL tends to be produced towards the end of thunderstorms.

We have also seen that winter thunderstorms differ from summer thunderstorms. Winter thunderstorms produce far fewer lightning strokes but a higher proportion of these are positive CG strokes (Rakov, 2003); and some of them are very strong. It appears that BL production is easier for these positive strokes than for the far more numerous negative ones. This intriguing correlation was first noted by Mark Stenhoff (1999). It is very important since it could give us some leverage to link the production probability of BL to the physical properties of particular lightning strokes.

Recently, there has been a first attempt to correlate these reports with data from lightning location systems in Central Europe. This was done by A. Keul, who has been collecting BL reports for more than 25 years, and G. Diendorfer of ALDIS (Keul and Diendorfer, 2018).

A number of cases were found where BL observation could be linked in space and time to a nearby lightning stroke. The authors conclude: "Contrary to detected lightning polarity frequencies, the assessed CG BL cases are associated with 54% positive and 46% negative strokes. The number of positive strokes over 100 kA associated with BL is remarkable, with 10 out of 19, but there are also frequent cases with negative strokes under 20 kA (11 out of 15)."

© Springer Nature Switzerland AG 2019
H. Boerner, *Ball Lightning*, https://doi.org/10.1007/978-3-030-20783-0_8

These findings add more weight to the hypothesis that BL production probability is higher for positive CG lightning than for other lightning. So what clues do we have now?

- Higher relative incidence of BL in winter thunderstorms that are rich in positive CG strokes.
- Higher incidence of BL towards the end of thunderstorms, where the positive CG strokes tend to occur.
- The recent analysis of Keul and Diendorfer.
- Cases where many BL objects were created by positive lightning.

We have not yet discussed the last point. Cases where many[1] BL objects were observed are much rarer than other cases. Fortunately, one of these is one of the best documented cases of BL sightings at all. The other cases vary considerably in the detail of the description; some of them are very detailed reports with some figures included, others contain only brief descriptions of the circumstances. I made the present selection by checking whether any information about the circumstances of the observation could be found. The full reports are given in the appendix.

Neuruppin 1994 (Case 1)

In Neuruppin, a small town north-west of Berlin, on 14 January 1994, there was a rather warm day for mid-winter. There were clouds but no rain, except maybe for a bit of light drizzle.

Until shortly after 17:00 local time (16:00 UTC), there was no indication that there might be a thunderstorm in Neuruppin. By that time, it was already completely dark. In 1994, Neuruppin still had a meteorological station with humans keeping the records, and that turns out to be very important in this case. The meteorologist on duty, Th. Hinz, described the event that happened so suddenly at this moment: "An exceptionally bright flash of light was seen originating from the north, followed by a very loud clap of thunder about 10 seconds later." Later, the staff of the station observed three more lightning discharges. Because the office had no windows facing north, the source of the light could not be seen directly, but all objects in view were brightly illuminated. The weather diary of the station notes: "Thunderstorm (light) N-E 16:06–16:28" (time in UTC).

---

[1] "Many" here means more than two.

The thunder must have been really loud, since a number of people compared it to bomb or artillery shell explosions, so it is no wonder that, soon after the first discharge, telephone calls were coming into the station, with people inquiring about the nature of the light and the noise. The director of the station, Donald Bäcker, collected the information the callers gave about their observations and he also put a request in the local newspaper for more eyewitness reports about this unusual event. However, he did not mention BL in this request. To his surprise, the reports did not only contain accounts of the primary flash, but often described luminous spheres above the rooftops or inside houses. He was at first inclined to dismiss these reports as rather far-fetched, but as more and more reports of a similar nature continued to come in from different people, he became convinced that they were real. In total, 34 reports were collected, including the identity of the observers and a brief transcript of their observations. After finishing the collection of reports, D. Bäcker published a summary in the local newspaper on 31 March, and in the supplement of the internal report of the German weather service. Unfortunately, the observation of large BL objects above the houses led him to the conclusion that the primary lightning had also been some sort of BL.

Meanwhile, two amateur scientists, Katja and Sven Näther, became intensely interested in the event. On their own initiative, they visited all the witnesses for interviews. With few exceptions, people were ready to give more information about their observations. The interviews took place in the 12 months following the event, so recollection of the observations would have been relatively fresh. The Näther couple went much further in investigating the observer reports than the meteorologists; they even took photos of some of the locations. The results of their investigations were summarized in a small booklet which they published themselves.

In these reports, five classes of phenomena can be distinguished:

- the primary lightning flash (the light and the loud thunder),
- corona discharges, some of them of extreme size,
- two large BL objects over the rooftops and over the lake,
- several small BL objects inside or around houses,
- a residual category which includes widespread damage to telephone systems and the like.

The spatial distribution of the observers' locations was centered at Neuruppin, and covered several square kilometers (see Fig. A.2 in the appendix). There are no reports from villages further east or west, but some observations have been made in parts of the town at the eastern shore of the lake. I first read the account on the internet after attending the BL conference in Antwerp in 1999 and then I contacted the management of the BLIDs lightning location service run by Siemens, which had been in operation in Germany since 1992, in order to obtain more information. The manager, Dr. Thern very kindly sent me the report of the thunderstorm activity on that day over a square region of side 40 km, centered on Neuruppin. I remember the moment I first looked at the data: there were only a few strokes recorded and the first fitted the time given by the meteorologists very well. Yet its strength was quite unusual, it was a positive CG stroke of 370 kA maximum current. I had to check twice to see whether the decimal point was really at that position before I could believe it. Such a maximum current lies at the very high end of the distribution for positive CG strokes. There can be no question that this positive lightning was the source of all the BL objects reported. There was simply no other lightning earlier that day in the vicinity of Neuruppin, and the other strokes were further away and also later. With the few lightning strokes in winter thunderstorms, it is much easier to establish a correlation between particular strokes and BL observations, whereas the high number and also sometimes high rate of strokes in summer thunderstorms often makes this impossible, like finding a needle in a haystack. The stroke cannot have been a perfectly normal positive CG stroke, because besides the enormous strength, only one observation was reported which could be related to this primary stroke. Furthermore, the distance from the strike point to Neuruppin is very unusual: the lightning location system placed it more than 5 km east of the town in a forest (see Fig. A.1 in the appendix). The lightning must have acted at a distance, even allowing for a larger location error than usual,[2] it must have been a long way from the locations of the BL observations.

The origin of these strong positive lightning strokes in winter could be the effect known as "wind shear". The wind speed higher up in the atmosphere is greater than close to the ground and the convective thunderstorm clouds in winter are less tall than their counterparts in summer. Therefore, the strong upper winds can blow away the positive charge in the topmost part of the thunderstorm cloud faster than the lower negative charge. The positive charge is then exposed to the ground and consequently one observes a bipolar pattern: downwind positive CG strokes and upwind the negative ones. Such a pattern also appears in the area around Neuruppin. The lightning detection system recorded a cluster of negative lightning flashes close to the town of Kyritz, about 30 km west of Neuruppin. In Kyritz, a very strong corona was observed; one observer, who also was meteorologist, compared it to the blue

---

[2] The accuracy of BLIDS at that time was probably 1–2 km (Heidler et al., 2004).

warning light of police cars illuminating the antennas on the rooftops, but no
BL objects of any kind were observed in Kyritz.

On this day, maritime air of polar origin was moving at a speed of about
90 km/h from the west–northwest over northern Germany. Could it be that
that the lower, negative end of the charge dipole caused these negative flashes,
whereas the positive charge, displaced in an easterly direction by the wind
shear, produced the predominantly positive flashes east of Neuruppin?

Unfortunately, although this explanation looks fine at first glance, it is
almost certainly wrong. The negative flashes around Kyritz were 18 minutes
later than the first positive flash at Neuruppin, so at that time the negative
charge would have been lagging even further in the west. It is more likely that
the charge structure of the thunderstorm cloud responsible for the strong
+CG stroke at Neuruppin was an inverted dipole, where the positive charge is
below the negative one. The last action of the cloud was a simultaneous discharge of a positive and a negative CG stroke, so this would fit the inverted
dipole hypothesis.

In addition to these two bipolar clusters the thunderstorm produced only
two other flashes in a square region of side 150 by 150 km centered on
Neuruppin.

In Neuruppin, a whole collection of different BL objects was observed.
There were two large ones outside, one standing still over the rooftops, the
other moving over the bridge which spans the lake. Smaller ones were seen
outside houses and also inside: two moving from outside into the houses, one
through a window—through which it left again—and the other entering
through an open window but passing through a closed curtain. The curtain
afterwards showed burn marks, but it was thrown away before the Näther
couple could investigate it. Two BL objects were seen inside houses from start
to end, both were very short-lived and didn't move at all. Another one, also
motionless, stood for a considerable time between an antenna and the railing
of the veranda. Yet another one was seen moving outside from the roof down
to a window and then up again. It was observed at rather close distance and
its shape was described as similar to a satellite dish, not a sphere. Others were
spherical or at least "egg-shaped". Critics often claim that BL is an ill-defined
phenomenon. In order to address this criticism, one can of course take the
range of behavior of the BL objects from Neuruppin and use them as a
description of the core phenomenon, since it is clear that all were due to the
same origin—the strong positive CG stroke—and they were most likely also
the same type of object. Of course, one could argue that the lightning stroke
may have produced several types of similar objects, but here Occam's razor
tells us that, since we have no indication that this was the case, we should

continue to work under the assumption that all the observed BL objects were of the same nature.

The number of observed BL objects was certainly greater than 11, since in a few cases people were not willing to be interviewed by the Näther couple, but from their earlier phone calls it was obvious that they had observed something similar. It is reasonable to assume that a considerable number of BL objects simply were not observed at all, so the number of objects must have been higher: the conditions for BL creation must have been excellent.

When I got the information from the lightning location system, I contacted both the Näther couple and Donald Bäcker, the meteorologist on duty, and we had a small meeting where we discussed the event and its interpretation. The result was published in a short paper (Bäcker et al., 2007).

Neuruppin is not only special because of the number of BL objects observed, it is also the only case I know where the information was collected by professionals *before* it became clear that BL had been observed. The action of the meteorologists was triggered only by the exceptional lightning. The situation is usually the other way round: people have observed BL and they contact somebody to get more information, or they are asked as part of an inquiry about BL observations. In Neuruppin, other phenomena, like corona discharge, were also reported, and this would normally have been lost from the reports.

Amiens 1884 (Case 5)

What happened in Amiens, in northern France, on Sunday, 24 February 1884, looks almost like a carbon copy of the event at Neuruppin. The day was relatively warm (10 °C), but there was a lot of rain and even hail. At 7:45 pm the only lightning of this thunderstorm suddenly struck the town. It was described as "terrible" and shocked many people because of its strength and because it came so unexpectedly. This event must have aroused the attention of a scientist, because soon afterwards, an account was published in a French journal (Decharme, 1884). The report states:

"Despite the fact that there was only a single lightning stroke in this thunderstorm, it hit in globular form at seven places, the most outlying ones being 1340 m distant." The report gives rather detailed descriptions of these seven observations and even a map of the town with the sites of the observations is included. All the BL objects were very small, and were compared to nuts and eggs. The report describes one of them entering the theater through a window during a performance, punching a small hole in the pane of 3 by 2.5 cm (a sketch of the hole is even included in the report), then passing a group of actors behind the scene, singing the trousers of one actor above the knee, and finally disappearing below the scene. Obviously, there was fear that it would start a fire, which is always very dangerous in places where many people are gathered, like a theater, so an investigation was immediately carried out to see whether there was any more damage done, but the object had disappeared without further trace.

Of course, one cannot be certain that this very strong lightning was a positive CG stroke, but singular strong positive lightning is known to occur in winter and the similarities with Neuruppin are striking. The small size of the BL objects is rather surprising, in Neuruppin there were some very large objects and some quite small ones, but in this case, all were of comparable size and quite small size. Brand lists this observation as Case 55.

Lamington 1922 (Case 3)

The report on the Lamington BL sightings is contained in the book called "Green Mountains" (O'Reilly, 1962) mentioned in the introduction. It was this account that started my interest in BL. Bernard O'Reilly was a very keen observer of nature, probably better than most physicists. His description is brief (Case 3), but it contains a lot of important details. The place where he observed the thunderstorm can be reconstructed exactly. He was on the way to a town with his pack horses, following the old trail that members of his family had built to improve access to the farm. He had passed an old hut ("humpy" in Strine) which had belonged to somebody called Luke and stood at the cliff's edge. Figure 8.1 shows a map of southern Queensland featuring the old pack track and a cliff marked "Lukes Bluff". A short distance back to the east two houses were marked on an old map, in one of which Bernard must have sought shelter.[3]

He describes the thunderstorm as exceptionally strong in terms of lightning rate, and also adds that the clouds were special: "… the clouds were higher than the usual storm and tinged with reddish brown, and as they advanced a constant rain of violet chain lightning fell on the undulating country below."

He describes very vividly how he found shelter in the empty hut and how the hut was hit by lightning. He only saw the two BL objects when he opened the shuttered window: "two balls of fire were drifting slowly past the humpy

---

[3] The coordinates are: –28.2186408,153.1163594 if you want to check it out on Google maps, or you can search for Luke's Bluff.

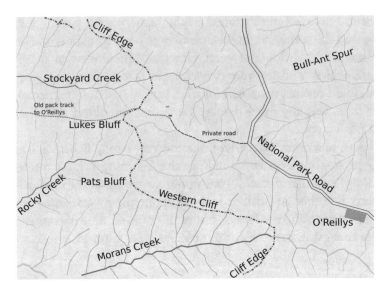

Fig. 8.1 Map of Lamington National Park, Queensland. The track Bernard was following with the horses is marked "old pack track to O'Reillys". Luke's hut was probably the house at the end of the private road. Drawing by the author

about fifteen feet from the ground; they were about the size and shape of a soccer football and were a deep glowing red like the coals of a burning iron-bark log; they drifted idly this way and that and it was the very uncertainty of their purpose which made them so terrifying."

The thunderstorm lasted for nearly an hour, and during this time, he observed a considerable number of BL objects: "At times I looked out; the horses were weathering it all right; always there were fireballs drifting; at times they exploded and the red light which flooded the humpy brought with it a wave of heat."

The sensation of heat could have been produced by infrared radiation, but also by a short burst of microwaves.

The production probability for BL must have been very high indeed, so it is of great interest to reconstruct the conditions that were responsible for this. Obviously, the updraft in the thunderstorm cloud must have been stronger than usual, leading to cloud tops that were significantly higher than for other thunderstorms. Moreover, the electrification process must have been exceptionally strong, the lightning rate being described as "like machine gun fire". Another clue is the color of the clouds: they were tinged with brown. The strong updraft may have brought dust or even smoke from bush fires into the cloud, and it is well known that smoke can considerably increase the lightning rate (Thornton et al., 2017). Bush fires in the west were observed earlier during the total solar eclipse, so it is not unlikely that some fires were also burning during this thunderstorm. Smoke from bush fires is also known to create a higher percentage of positive CG lightning (Rakov, 2003). We can conclude that there was an exceptionally high rate of lightning and probably a high percentage of positive lightning, and this coupled with the exposed position of the hut on a flat area but close to a cliff may have been responsible for the exceptional BL creation rate.

How reliable is this exceptional report? The book was written in 1939, and the event took place in 1922, so how much could he have remembered? The witness was clearly a very good observer of nature and, since the situation must have impressed him considerably, it seems likely that he would be able to clearly recall details even many years later. The report of the thunderstorm is only a small paragraph in the whole book, so it is very unlikely that he made it up; there was simply nothing to be gained from putting it in a report about a settler's life on the Lamington plateau in the McPherson mountain range.

The description of the lightning strike to the hut is so detailed that it can be used to falsify theories of alternate BL explanations, as we shall see in the next chapter.

A similar case is described by Brand (case 36 in his book, not included in the appendix) which also happened during a very violent thunderstorm, but less information is available concerning the circumstances. In this case 25–30 BL objects were seen within half an hour.

Santa Maria 1902 (Case 4)

This BL report is special because it is not connected with normal thunderstorms, but with volcanic lightning. The report is short, not very detailed, and only second hand, but it mentions both corona discharge and BL during the eruption of the Santa Maria volcano in Guatemala in 1902. This eruption was the second strongest in the twentieth century, with a volcanic explosion index (VEI) of 6 out of 8, ejecting about 5.5 km^3 of magma.

On 25 October, at the climax of the eruption, in San Cristobal Cucho, 30 km north-west of the volcano and downwind, strong electrical effects were observed that were obviously corona plus a lot of BL. The report says: "the electricity went out of the cloths and the body of the people and of the houses. Ball lightning everywhere, exploding with a hollow sound without doing damage." In this region there was a heavy ash fall (a total of 1 m), so the situation must have been truly chaotic. During the previous day a lot of lightning had been seen in the eruption column: "24 October, 8:00 pm: From San

Felipe one could see a giant plume of black ash with numerous fierce twisters crossed by thousands of lightning bolts and curved lines of red light." On 25 October, there must have been lightning, but there are no reports of CG lightning in the vicinity of Cucho, so it would appear that the BL was created by the electric field at ground level, which also produced the corona discharges. The polarity was probably as in a positive CG stroke, since the ash column at this distance must have been positively charged.

An eruption like this one appears to offer the opportunity to observe BL in a semi-controlled way: today, there is usually good monitoring of dangerous volcanoes, and when an eruption is imminent, many geologists and other interested people flock to these spots to make videos of the eruption. It can be expected that BL will not go unrecorded, even when the ash-fall is severe.

Martinique 1891 (Brand Case 103)

The Martinique report links severe hurricanes, in this case on 18th of August 1891, with BL. This is unusual since lightning is not normally so frequent in hurricanes. The report states: "Ball lightning was seen everywhere during the hurricane. Country people talked of fireballs, which flew through the air for several minutes while making a crackling noise and burst at about 50 cm above the ground."

The lightning appears to have been concentrated around the route taken by the eye of the hurricane: "The electrical discharges followed one another incessantly; lightning flashed continuously: it increased and decreased in intensity and number before and after the passage of the storm center". As mentioned above, most hurricanes produce very little lightning, but there are exceptions. Lightning location systems show that episodes of strengthening and also weakening tend to go along with more lightning and especially with a high percentage of positive CG lightning (Thomas et al., 2010). During an episode of the very strong hurricane Matthew in 2016, many sprites were observed from a distance over the storm.[4] Sprites are almost without exception associated with strong, positive CG lightning, so does this prove that at least hurricane Matthew created a large number of these strokes? And could this also have been the case in Martinique? Well, this can be concluded only with a large grain of salt: there are rare cases where strong negative lightning also creates sprites, and very strong negative CG lightning occurs almost exclusively over the ocean.

---

[4] Search the web for "hurricane Matthew sprite" to see a number of photos.

The important question is now: can we learn anything from this correlation between BL and positive lightning, which could shed some light on the nature of these objects?

Summary

- There is now clear evidence that the incidence of BL is more closely correlated with positive CG lightning than with negative CG lightning.
- The Neuruppin case stands out because it is especially well documented, and we know exactly what created the large number of BL objects: the correlation of BL creation with the extremely strong positive CG stroke is unambiguous.
- The BL objects created in Neuruppin show almost all aspects of BL behavior reported in other cases: large stationary and moving BL objects outside, small BL objects created inside houses, BL objects passing through windows and curtains, etc.
- Other cases with a copious production of BL objects can also be correlated with positive CG lighting, but with less reliability.

9

The Link Between Lightning Physics and Ball Lightning

We have now seen that positive CG lightning is more frequent in winter, which correlates well with the observation that the incidence of BL is higher in winter than in summer. We have also seen that some of the more spectacular cases, especially with multiple BL production, are associated with positive lightning. This correlation is in itself an interesting and surprising finding, but can we also learn something about the physical nature of BL from it?

Clearly, we need to identify the differences in the conditions created by positive lightning and negative lightning, but we know this already from the previous chapter on lightning physics: the attachment processes of negative CG and positive CG strokes are different. Because of the change in polarity, positive CG strokes produce a negative space charge above ground and not a positive one as for negative CG lightning. The connecting negative streamers which are finally created need a higher electric field than positive streamers in order to start and propagate, roughly a factor of 2–3 stronger (MacGorman and Rust, 1998). This means the leaders are suppressed, and when they start, they are late and the electric field at that time has a much higher energy density than when positive connecting streamers start to grow. The energy density of the electric field goes as the square of the field strength, so 4–9 times more energy per volume is available at ground level when a positive leader comes close from above (Fig. 9.2). The streamers basically reduce the electric field in a region around their stem; because of their high conductivity they are almost at ground potential and the energy difference in the field is converted into the ionization of the streamer plasma, the heating of the plasma, and a small amount also into light emission. Figure 9.1 shows the situation for streamers growing from the ground.

© Springer Nature Switzerland AG 2019
H. Boerner, *Ball Lightning*, https://doi.org/10.1007/978-3-030-20783-0_9

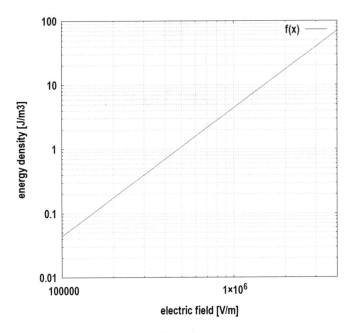

Fig. 9.2 Energy density of the electric field. Diagram by the author

Any other process competing for energy with the connecting streamers will profit when streamer creation is delayed, because a larger amount of energy remains available. If we now assume that the creation of BL objects is correlated with the energy density of the electric field, it is reasonable to expect positive CG to be more likely to produce BL than negative CG. Under this hypothesis, BL creation competes with the streamers for energy and only when enough energy is available locally to start the creation process is there any chance that such an object can be produced.

Is this a reasonable proposition? In fact, the difference in field strength and energy density for streamer growth is known from textbooks (MacGorman and Rust, 1998). The resulting difference in the attachment process of negative CG and positive CG has been observed (Wang et al., 2016). The only assumption is that the creation probability of BL and the energy density of the electric field are linked: the more energy, the higher the probability.

Of course, the electric field is only a weak source of energy. It cannot supply enough energy for very energetic objects. Figure 9.2 shows a diagram of energy density versus field. However, for the BL objects that are typically observed indoors, the energy may be sufficient.

We now have the following differences between positive and negative CG lightning:

Fig. 9.1 Upward connecting streamer: the left streamer didn't quite make it, while the connection with the stepped leader was indeed made on the right. With kind permission, copyright Hank Schyma, alias "Pecos Hank": https://pecoshank.com

- Positive CG lightning produces negative space charge above ground, rather than positive.
- Negative streamer development is suppressed because they require a higher field strength than positive streamers.
- The high electric field strength leads to a much higher energy density in the electric field above ground.

If these are the ingredients for good BL production, we can understand why tower-initiated lightning and rocket-triggered lightning with a grounded wire have such little success in producing BL: in these cases, the discharge starts upwards and the upward growing leader is basically at ground potential. This therefore reduces the electric field around the tower, and neither a space charge nor any other streamers can develop there. The wire connecting the rocket to the ground also suppresses any high field around ground level.

Negative CG lightning, on the other hand, poses a problem: it produces neither a negative space charge, nor such a high electric field above ground, yet we know that such lightning is perfectly capable of creating impressive BL objects, as in Devon, where a 15 kA negative lightning stroke hit a neighboring house (Case 15). This was rather run-of-the-mill negative lightning, and not one of the especially strong positive strokes. Obviously, we are missing something here. Either the hypothesis that negative space charge and a high electric field are important ingredients for making BL is wrong or we have overlooked something important. At this point, we need the input from a theoretical model in order to get some guidance for the interpretation. But before we come to the discussion of the many BL theories, let us take a look at the arguments of skeptics who claim that BL is just a figment of our imagination, as in Macbeth: "… a dagger of the mind, a false creation".

Summary

- Positive CG lightning correlates better with BL production than negative CG lightning.
- Differences in the attachment process of −CG and +CG lightning offer a good hypothesis for the higher probability of BL generation by +CG.
- +CG lightning produces a negative space charge above ground.
- +CG lightning produces higher electric fields around the hit point than −CG lightning, without creating streamers that compete with BL for energy.
- The energy density of the electric field around +CG hit points can be up to a factor of 10 higher than for −CG without creating discharges, giving competing processes like BL objects a better chance for creation.
- This hypothesis can explain why tower-initiated lightning and rocket-triggered lightning with grounded wires have such a low probability of producing BL.
- The observation that rather weak −CG lightning can produce BL objects points to the fact that this explanation is not yet sufficient to fully specify optimal conditions for BL creation.

10

Some People Just Won't Believe It: The Skeptic's View

History of BL Skepticism

Many scientists have always been skeptical concerning BL. The reported properties simply didn't fit into any known theoretical framework. This was not so obvious when physics itself was less advanced, but since the 1950s, when the structure of matter had been unraveled and plasma physics had been developed, BL simply had no place in the context of known and understood phenomena. Currently, only very few scientists are actively working on this subject, and most regard BL as a pseudo-phenomenon or chimera (Campbell, 2008). Historically, misidentification of BL with other natural phenomena has been proposed by all skeptics. As Brand's analysis and Stenhoff's analysis show, BL reports must be scrutinized and checked carefully for the credibility of evidence, but there remain a considerable number of reports that are very detailed and from sources that cannot be easily dismissed as unreliable. Therefore, alternative explanations centering around misperceptions or illusions have been developed, the earliest being afterimages created by the exposure of the retina to strong light, for example from a lightning channel. Afterimages are usually observed when one unintentionally looks at the Sun: a spot—either bright or dark—is observed for several seconds because the retina is not responding normally at the spot where the image of the Sun was focused. The afterimage fades away within a few seconds, but eye movement may create the illusion that an object is moving in the field of view. For lightning to produce an afterimage in the form of a round spot, the channel has to be seen end-on, otherwise one would perceive a jagged line as an afterimage and no glowing globe. Since it is very rare to see a lightning channel end-on,

© Springer Nature Switzerland AG 2019
H. Boerner, *Ball Lightning*, https://doi.org/10.1007/978-3-030-20783-0_10

the conditions for such an illusion are rarely fulfilled. In addition, many observers of BL objects could never have seen the initial linear lightning, so in these cases such an explanation is not possible. All the BL observations in Neuruppin (appendix case 1) fall in that category. More recently, several different, but related phenomena have been proposed as the source of BL illusions: stimulation of the brain or retina by magnetic or electric fields due to a nearby lightning stroke. Before discussing these stimulation hypotheses in more detail, we will have a look at the arguments and strategies of three well-known BL skeptics.[1]

Humphreys

In 1936, Humphreys published a very skeptical paper on BL (Humphreys, 1936). Apparently, he had previously thought that BL could not be an illusion, but after collecting and analyzing 280 BL reports he changed his opinion: "Not one of these many accounts is unmistakably of what ball lightning is supposed to be, that is, a leisurely moving and approximately spherical body of gas, electrons, or what not, luminous by virtue of its electrical state or condition." He dismisses all reports except "maybe, two or three" as being due to afterimages by brilliant flashes, broken discharge paths, meteorites, will-o'-the-wisps, falling molten metal, lightning seen end-on, and brush discharges. Reading his paper one gets the distinct impression that, in trying to be critical about the reports, he has been carried away by his imagination. In one case, he explained a report to be quite clearly a typical observation of a will-o'-the-wisp: "this is a most excellent description of one of the recognized varieties of the will-o'-the-wisp, namely, an owl on a hunting flight and covered with fox fire from a decaying hollow tree where it had spent the day." He is thus suggesting that a bioluminescent owl is a satisfactory explanation of a BL report. Fox fire[2] is an old word for the very faint luminosity emitted by several species of fungi living on decaying wood. Such luminosity can only be seen by a fully dark-adapted eye on a completely dark night, and can be photographed only by a long-time exposure, so it would be really far-fetched to associate it with BL. Trying to explain BL with such an ill-defined and poorly understood phenomenon is simply ridiculous.

---

[1] Recommended books from two serious skeptics: M. A. Rothmann: A Physicist's Guide to Skepticism, Prometheus Books and C. Sagan, The Demon-Haunted World, Headline Book Publishing.

[2] Fox fire may be derived from "faux", French for false; so, it means "false fire". This may also be the origin of the name of "Fox News" in the US.

Another example shows how hard he tries to explain away BL observations, in this case by Bernard Loeb, the professor of physics who was an authority of electrical discharges in gas: "As I looked out of the window I noticed a ball of what I would now describe as the color of active nitrogen or possibly slightly darker, as it seemed to me, descending from somewhat the direction of the neighbor's house in a light graceful curve. Its diameter appeared to be about double that of the toy balloons which one sees and its motion through the air was quite analogous to the motion of the type of air-inflated balloons which are used so frequently in modern dinner parties. It had a translatory motion in my direction and seemed to descend an inclined plane from the approximate location mentioned. It appeared to strike the lawn, bounced slightly once and then disappeared." Loeb's full account is given in the appendix. Humphreys now uses three different arguments to explain this observation:

"I fully concur in the supposition that this phenomenon was started by an intense glow, or brush discharge, but strongly suspect that its graceful fall was the familiar travel of an after image, and its bouncing just what any "born" physicist would expect a ball to do and, if it were an after image, "see" it do. The crash of thunder distracted attention and the ball, doubtless already fading, was lost to sight."Only the first one, the glow discharge, is credible, but he fails to explain why a relatively weak glow discharge could have produced an after image, and he then needs to explain its observed fall to the ground. The bouncing on the lawn he cannot explain at all, so he must claim that Loeb, as a born scientist, was just expecting this to happen. Obviously, this ad-hoc hypothesis is as far-fetched as can be, strongly violating Occam's principle of simplicity.

In the end, Humphreys is not saying that there is no BL: "Is there, then, no such thing as ball lightning? I don't know. I only know that many things have been called ball lightning that were something else." This is certainly true, but he clearly overshoots the aim by a large margin by trying to be too critical. Brand was also very critical with respect to BL reports, so he had to dismiss two thirds of the original accounts, but in the end he still had enough credible ones to use in his analysis.

Campbell

More recently, Steuart Campbell has been one of the more vocal skeptics concerning BL (Campbell, 1993, 2008). Once again, as was the case with Humphreys, he appears originally to have considered BL to be a phenomenon of significance, but then he came to question BL as an independent phenomenon, and found that it could not be explained by already known effects. Being a professional skeptic, member of the Edinburgh secular society,[3] he published several papers concerning pseudoscience, etc. It seems that he also fell victim to the urge to criticize over excessively; in Campbell (1993) he not only heavily criticizes a BL paper, but also makes scathing remarks about human-induced climate change. Being a climate skeptic, disputing the greenhouse effect of CO_2, seriously undermines his credibility. In Campbell (2008) he claims that BL properties are all over the place: "In other words, the phenomenon exhibits no consistent characteristics and appears to be all things to all observers." In other words, he ignores the fact that many statistical analyses that have been perform since Brand's time, demonstrating that BL observations are consistent with a core phenomenon with well-defined characteristics, as clearly shown by Keul and Stummer (2002). He also states with respect to BL reports: "Consequently, genuine anecdotal reports of BL must be regarded with suspicion. Observers are mostly unaware of the distortions involved in perception and memory. Worse still, asking people if they have seen BL begs the question of its existence and ignores their inability to distinguish it (if it exists) from other phenomena." He should be aware of the selection process performed by Brand which deals with the unreliability of BL reports. Furthermore, the paper on the Neuruppin event was published in 2007 (Bäcker et al., 2007), clearly showing that the information was not collected by asking for BL events. And several BL photos were available by that time. The minimum requirement one must ask from a skeptic is to review alternatives to his or her propositions. Campbell also asks us to consider the so-called "null hypothesis". We will see below what that means.

Berger

Berger was an eminent lightning researcher pioneering tower-based lightning studies of lightning currents and characteristics.

He used data from oscilloscope traces of currents measured using shunts installed at the tops of two 90 m high towers on the summit of Monte San Salvatore in Lugano, Switzerland. The towers were not very high, but because the mountain height of 915 m above sea level contributed to the electric field at the tower tops, the effective height of the tower was several hundred meters. As a result, most of the lightning strikes to the towers were of the upward-

---

[3] From the English Wikipedia: https://en.wikipedia.org/wiki/Steuart_Campbell

moving type. His research was very influential, and indeed the results are still being used for lightning protection and in lightning research.

In 1973, he wrote an oft-quoted paper commenting on his 30 years of lightning research and BL (Berger, 1973). His skepticism concerning BL comes from two sources: during his study of lightning he never saw a BL object, and several BL reports he investigated could readily be explained by other phenomena. He concluded that BL had never been photographed up to the year 1973 and that the many reports of BL observations were in the first place produced by the effect of the lightning and its consequences on the subjective perception of the observer. Specifically, he mentions afterimages, quoting Humphreys. He also proposed to ignore older reports because they could not be analyzed further and asked for new reports to be scrutinized in field investigations by high voltage engineers and physicists.

There are several problems with his conclusions. First, the fact that he had never observed BL—even though he had seen thousands of lightning flashes—cannot be used as proof that BL does not exist. In such a situation, where nothing was observed, the only thing one can derive is an upper limit on the production probability of BL under the particular experimental conditions. "I haven't seen it, so it doesn't exist" is not a valid conclusion. Absence of evidence is not evidence of absence. Furthermore, almost all the lightning he observed was of the upward-moving type, and as explained above, in these circumstances the electric field strength at and around the tower can never reach the high values one obtains on a flat terrain when the leader comes down. If BL production is correlated with the energy density of the electric field created by the leader, he was working under particularly unfavorable conditions. His proposition to ignore older BL reports is also not useful. The reports in Brand's book are all from a time where there were no high voltage lines around, so flashes at these lines are excluded as explanations. I suspect that he never had the opportunity to read Brand's book, because he remarks that it was long out of print and he does not quote its content exactly.

Stimulation of the Brain: Phosphenes and Friends

Soon after friction machines became available for producing electricity in the eighteenth century, it was detected that electric shocks could lead to contractions of the muscles. It became a popular game to "electrify" several people that had clasped hands, sending them all jumping high. More recently, it was found that the brain could also be directly stimulated by electrodes, but this required drilling holes into the skull, so these were rather grizzly experiments.

Today, brain stimulation can be achieved without cutting the head open: so-called trans-cranial[4] magnetic stimulation uses coils with strong, time-varying magnetic fields to induce currents in the brain that provoke reactions. If the visual cortex, the region of the brain that processes visual information from the eyes, is stimulated, the impression of visual effects like luminous objects can be provoked in the subject (Kammer et al., 2005). These illusions are called magnetophosphenes, or phosphenes for short. This method works because magnetic fields can penetrate the head, whereas electric fields are better shielded by the conductivity of normal tissue. It was first proposed that, akin to the stimulation by coils attached to the head, the magnetic field of a nearby lightning strike could stimulate the visual cortex, leading to a visual sensation that simulates what is then referred to as BL. It turned out that the initial calculation overestimated the effect on the brain (Peer et al., 2010), so the hypothesis had to be changed: then a direct stimulation of the retina was proposed, occurring at lower magnetic fields than the stimulation of the cortex (Peer et al., 2010). This hypothesis also encounters difficulties, since it requires repetitive stimulation by several consecutive strokes in one flash. We have seen above that positive CG lightning is correlated with several spectacular BL observations, and a positive CG consists almost exclusively of just one stroke. In addition, in the case of Neuruppin, the distance between the point of the positive stroke and the BL observations was on the order of several kilometers, so no magnetic field of appreciable strength could have been present.

A very thorough analysis of the magnetic phosphene hypothesis was published in 2008 by A. Keul, P. Sauseng, and G. Diendorfer (Keul et al., 2008). It reached the following conclusion:

"Summing up the relevant points of the authors' first empirical assessment of the BL EM hallucination hypothesis, its explanatory power for BL observations is not high, because:

- the magnetic stimulation level of 0.01 T (retina) versus 1 T (brain tissue) is not reached in cases of nearby average cloud-to-ground flashes;
- only some of the lightning flashes will produce subsequent strokes, i.e., "needle impulses" necessary for repetitive phosphene generation;
- magnetic stimulation by lightning can only explain simultaneous visual perceptions with a maximum duration of one second;
- short-range lightning will not produce local field differences relevant for cortical magnetophosphene effects, but a rather homogeneous field;

---

[4] Trans-cranial means "through the skull".

- there is a phenomenological mismatch between the optical appearance of possible EM phosphene hallucinations and BL;
- phosphenes are not commonly reported as medical lightning strike symptoms.

Therefore, we reach a negative conclusion for the EM BL hypothesis."

Unfortunately, the idea of proposing illusions due to the effect of lightning on the brain as an explanation for BL is very popular in scientific circles. In Stenhoff's book, a multiple BL observation is mentioned which is sometimes used as evidence supporting the brain stimulation hypothesis. The report from a certain Mr. Swithenbank is as follows:

"The event occurred when a house party was in progress and about forty people were present in the house. Those of us in the kitchen saw the ball in the middle of a group of several people, roughly looking in the direction we were looking when the lightning struck. The people in the lounge similarly saw the ball near the fireplace. The girl in the bath saw it towards the window. I feel sure it could not wander all around the house, and nobody has ever seen two at once to my knowledge… As a physicist, I interpret ball lightning as the effect of a very strong electromagnetic pulse on the brain." This personal communication to Mark Stenhoff is a bad example of a prejudiced scientist jumping to a conclusion without properly analyzing the observations. Here are some questions that he should have answered: What was seen by the various witnesses? Was there agreement in their observations? What was the shape, color, and motion of the objects? Were other observations made? Where did the lightning strike? Was it close or far away? What would be an estimate of the magnetic/electric field produced by the stroke? Would that have been sufficient to explain the observations? Are there alternative explanations? In addition, "I feel sure" is no replacement for a proper investigation, and "nobody has ever seen two at once", no, that is not correct, two BL objects were seen at Neuruppin and at Lamington, and other cases have already been mentioned by Brand. So, this case is an example of a lost opportunity, it should not be quoted as an example supporting the idea of the brain stimulation hypothesis.

The latest addition to the list of effects that could simulate a BL observation is the proposal that certain epileptic seizures can be triggered by lightning that is not necessarily close by (Cooray and Cooray, 2008). Some of these phosphenes do indeed resemble what people have reported as BL (Panayiotopoulos, 1999), but others are quite different. The authors proposing this hypothesis are therefore cautious in their claims:

"The possibility that some of the ball lightning reports are contaminated by the hallucinations caused by occipital seizures should be taken into account in future studies of ball lightning. It may help to separate the real physical facts from the hallucinations, paving the way for future progress in ball lightning research."In other words, they do not claim that all BL reports are due to this illusion, but only some of them. This hypothesis is also presented in a book on lightning physics by one of the authors (Cooray, 2015).

The "Null" Hypothesis

Other skeptics are not so cautious, and often adopt a much more radical position by asking whether the "null" hypothesis has been considered by BL researchers.[5] This states that all BL observations without exception are due to other phenomena like illusions or misinterpretation of natural phenomena. This is a sweeping claim and thus offers ample room for falsification: in principle, only one counterexample is needed to render the proposition false, and we can now produce not one, but several: we have seen several photos of BL objects, and the well-documented case of Neuruppin presents us with no fewer than 11 BL objects, spanning a considerable range of the properties reported for BL. The "null" hypothesis is often advanced by researchers who are deeply worried because they feel there is no way to understand BL as a natural phenomenon, but as we shall see soon, they are most likely wrong.

To end this discussion, I would like to reconsider the description by Bernard O'Reilly of his BL observation at Lamington (Fig. 10.1).

He unpacked the horses and then entered the hut to seek shelter. Then lightning struck the hut (a), the effects on his body were nearly as strong as by a Taser: he "went in a heap" (b). Then he got up, opened the shutter (c), and only then saw two BL objects (d), not before, as one would expect from the brain stimulation hypothesis. The only way to deal with this report as a hard-core skeptic is to question the reliability of the observer, but this is then a last resort that can basically be used to try to kill any account by any witness.

Before we close this chapter, I would like to make a remark on the latest BL comment I could find, from the skeptic Brian Dunning.[6] He runs a respectable website where he debunks pseudoscience, conspiracy theories, esoteric beliefs, and the like. He also thinks that BL falls into the realm of pseudoscience, so he lists four arguments against its existence as a physical entity:

---

[5] In statistics, the term "null hypothesis" is used differently.
[6] Dunning, B. "Ball Lightning." *Skeptoid Podcast*. Skeptoid Media, 9 Feb 2010. Web. 5 Jan 2019. <http://skeptoid.com/episodes/4192>

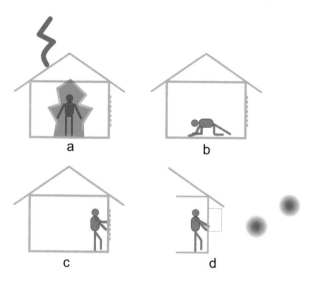

Fig. 10.1   Sketch of Bernard's account. Drawing by the author

- "Ball lightning is not reproducible in the laboratory. All known forms of electrical discharge are."
- "There is no standard description of what ball lightning looks like or how it behaves. Reports of its color, its size, its speed, its sound, the conditions under which it appears, its behavior, its shape, and its duration are all over the map."
- "Not a single photograph or video of ball lightning exists that is considered reliable and not otherwise explainable."
- "Electromagnetic theory makes no prediction that anything like ball lightning need exist. It does predict all know forms of electrical discharge."

The argument concerning the missing experimental evidence is certainly correct, but we have seen that the next two are wrong. There is a stable description of this phenomenon over centuries and there are now both photos and videos showing these objects. The last argument is particularly interesting. It basically consists of two separate parts: it assumes that BL is due to electromagnetic interactions and it states that electromagnetic theory does not predict it, or in other words, it cannot exist. This is a very sweeping claim, going much further than the statement that a theory has not yet been found. We shall see in the next chapter, when we look at BL theories, whether this strong statement can be upheld.[7]

Summary

- Lightning researchers like Berger operated and operate under experimental conditions that are very unfavorable for BL creation: the peak electric fields around lightning towers are always quite low compared to the situation in Neuruppin, for example.
- Only upper limits of BL production can be given; the conclusion that it does not exist because it has not been seen under certain experimental conditions is logically impossible.
- Alternative explanations for BL observations from skeptics focus on two lines of thought: misinterpretation of natural phenomena and lightning-induced hallucinations.
- In many cases, the misinterpretation of other phenomena can be excluded, for example, in the Neuruppin case.
- Lightning-induced hallucinations either require unrealistically high magnetic fields from the lightning return stroke or other characteristics like stroke repetition frequencies that have never been observed or can be excluded in cases like Neuruppin.
- Observations like the Lamington case demonstrate that strong physiological effects by very close lightning strokes do not necessarily lead to BL illusions.
- The "null hypothesis" can be falsified on the basis of photographic evidence and on well-documented cases like Neuruppin.

---

[7] Please don't get me wrong: skeptics like B. Dunning are doing an important job in a world that is ever more swamped with fake science and the like, but it seems that some of them have lost faith in humanity, so they are throwing the baby out with the bathwater by being too critical—at least in the case of BL.

11

Ball Lightning Theories

Status of Ball Lightning Theories

There are now probably more than 200 different theories of BL, but so far there is no consensus on the validity of any of them. In fact, most scientists think that none of these theoretical models is capable of explaining the physical nature of these objects. This is somewhat surprising concerning the effort that has gone into these models, but as Stenhoff explains in his book, most theories are critically flawed. He states that they simply do not describe the observational evidence: some predict behavior that has never been observed, whereas others are at odds with well-documented behavior by these objects. Some focus only on a few reports and ignore the evidence from other observations. It appears that physicists are also prone to do some cherry-picking when it comes to reports on BL.

Stenhoff remarks that there has only been one recent attempt to correlate observations with the theoretical models.[1] This is very surprising, because comparing theory to observation is the foundation of natural science and one should therefore expect any BL model to have been tested according to this procedure. Since this crucial check of theory against observation has obviously not been done sufficiently well yet, let us now take a look, and see if we can decide which theoretical models fit the facts.

In setting out to do such a test we need to define the starting point: should we require one model to fit all the observational evidence or should we allow several theories? It has often been stated that there may be more than one

---

[1] The actual report could not be obtained by the author: Hubert, P. (1996) Novelle enquête sur la foudre en boule—analyse et discussion de résultats, Rapport PH/SC/96001, Centre d'Etudes Nucleaires, Saclay.

© Springer Nature Switzerland AG 2019
H. Boerner, *Ball Lightning*, https://doi.org/10.1007/978-3-030-20783-0_11

phenomenon behind the BL observations, so would it not be natural to start with several models, covering different aspects of the reports? This is, in fact, a very bad idea, because it makes it almost impossible to eliminate theories from consideration. If a certain property of BL cannot be explained by one particular theory, one can simply claim that this is just because it is due to a different type of BL. In this way, one cannot reduce the number of candidate theories, while that is the very aim of this procedure. We will therefore start with the assumption that there is only one type of BL and that there is also only one way of producing it. On the basis of this strong requirement, we can eliminate all theoretical models that are not capable of explaining the observed modes of creation and the observed properties. We shall then see what theories remain. Only if none remains should we relax the requirement and consider the hypothesis that there is more than one type of BL. This procedure - starting with the simplest assumption and make it more complicated only when necessary - is, in fact, another application of Occam's razor.

Properties of Ball Lightning

The properties of BL have been discussed in depth above, but just as a reminder, they are reproduced here in a short list:

- The correlation with thunderstorms, especially with linear cloud-ground strokes (and preferably positive CG strokes). Creation at a distance from the lightning channel.
- Their stability for seconds, up to minutes
- Their ability to store energy, usually only a small quantity, but sometimes surprising amounts.
- Their stable spherical shape in more than 80% of cases, but also ellipsoids, toroids, or even irregular shapes
- The color and structure of their visible surface: sometimes smooth and quiet, sometimes active and sparkling, some BL surfaces are almost transparent, others are opaque.
- Their motion in open space, sometimes as if steered by some controlling agent.
- Their passage through glass window panes or other dielectric materials, and sometimes also through metal screens.
- Their appearance in closed spaces like rooms or aircraft.
- Their explosive or quiet modes of decay.

All of these properties present difficulties for the theoretical models, but three of them are especially hard to meet. The first "killer" criterion is the fact that BL objects can be created far away from the channel of a linear lightning. The lightning channel would be an ideal source of energy and plasma for the BL object, so many models assume that this is its source, but there is ample evidence that this is often not the case. The second extremely hard criterion is the passage through dielectric objects like window glass or wooden panels, also, a well-documented fact. The last one is their creation within closed spaces or even inside metal aircraft, which form Faraday cages. All these criteria are well supported by observations.

Classes of Models

As mentioned above, there are probably more than 200 different BL theories. It would be neither entertaining nor useful to check each of these in detail, so we consider only theories developed in or after the 1950s, and group them into five broad classes. Left out are the more exotic and obviously unrealistic ones like antimatter meteorites or magnetic monopoles. Readers interested in more detailed information can find this in Stenhoff's book, where he presents an almost complete list of the individual theories. The five classes considered here are:

- chemicals models, including burning substances,
- electric discharge models (DC),
- plasma structures, including filamentary ones,
- microwave bubbles,
- electromagnetic interference models.

Let's now take a closer look at each of these.

Chemical Models

Chemical models assume that BL objects contain reactive, burning substances like gases, or even a fine network of solid materials that is created by linear lightning strikes to the ground. These chemical reactions are assumed to proceed slowly, providing the energy for the BL in a more or less continuous way. Since chemically reactive substances have a high energy content, several researchers have found this class of models attractive. Nevertheless, they are plagued by several difficulties. It is hard to see why the luminosity from such chemical processes of reactive gases should be restricted to a spherical region alone, when lightning produces these reactive species all along the lightning channel. The substances created in the soil where the lightning hits would also need to form a ball, so a surface tension has to be invoked in an ad-hoc fashion for the burning blob.

Low-temperature burning of hydrocarbons has also been proposed as an explanation for BL, but this fails to explain where the rather high concentrations of these combustible substances come from, and why this correlates with thunderstorms. Another group of models claim that direct hits of linear lightning can create burning substances, e.g. silicon, if lightning hits quartz sand mixed with carbon. The high temperature could then reduce quartz to silicon, which would form burning globes. Other models explain BL with burning fractal structures of nanoparticles. There is no need to investigate any of these models in detail, since all of them fail one criterion: the burning matter claimed to form the BL object could never pass through a glass pane. Moreover, the models linking BL to the matter evaporated at the hit point of the linear lightning cannot explain their appearance far away from this lightning. We can therefore safely conclude that none of the chemical models can explain BL.

Electrical Discharge Models

Some models are centered around the idea that BL is just an unusual form of pulsating electrical discharge in air, like a glow or corona discharge (Lowke, 1996). Since this is an alternating current discharge, it could pass dielectric objects like glass panes. Nevertheless, such a discharge needs a current source and a strong electric field, and that will not be present in aircraft. It is also unclear how this discharge could move around for a long time, especially in houses. The required electrical fields have not been observed to exist for such long times.

Plasma Structure Models

The visible part of BL objects is of course a plasma which is radiating light, so many models have been developed around the idea that BL is basically a plasma structure, a plasmoid, preferably created directly from the lightning channel. The lightning channel is a copious source of high-temperature plasma and would also be capable of supplying a lot of energy. Unfortunately, this mode of BL creation cannot account for the numerous observations where BL objects were created far away from any lightning channel, for example in the well-documented case of Neuruppin. A plasma structure could also be created by different processes, but these models have other, serious deficiencies: the energy content of a plasmoid is limited by the so-called virial theorem, the electrons and ions forming the plasma would rapidly recombine to form neutral atoms and molecules, and plasma cannot be stable in such a well-defined shape for sufficiently long times.

Under atmospheric conditions, the ions and electrons of the plasma will recombine into neutral species within a very short time (about 100 ns), much shorter than the observed lifetime of BL objects. Some scientists have proposed that this recombination could be delayed by water vapor in the air, but why should the plasma of the lightning channel recombine within microseconds, whereas the plasma in the BL object takes many seconds to recombine?

Another serious problem is the notorious difficulty involved in confining plasma without using material walls. Since the 1950s, efforts have been underway to use a hot plasma of heavy hydrogen (deuterium), compressed by means of magnetic fields, in order to generate energy from nuclear fusion. It has been extremely difficult to produce a plasma of the required density and temperature because of the numerous instabilities encountered. So far, no fusion reactor has been created with a net production of energy. An experimental reactor called ITER (International Thermonuclear Experimental Reactor) is currently under construction in France. It is scheduled to start initial experiments in 2025 and finally to work with a deuterium-tritium plasma in 2035. If it is so difficult to enclose and contain a fusion plasma using magnetic fields, why should a plasmoid exhibit such a long lifetime? A more recent model postulates that the BL plasmoid consists of a knot of plasma filaments carrying a current that heats the plasma (Ranada et al., 2000). Such a tangle is supposed to be a more stable configuration because of topological constraints that block the unraveling of the knotted structure.

All these models, including the magnetic knot model, fail to account for the passage through glass windows. During the passage, all structures of the plasmoid would be destroyed because matter cannot pass the glass pane. The plasmoid should be uncharged, which means that it should contain about equal amounts of positive and negative charges. Electrons and ions would be hitting the outer surface of the glass, recombining and heating it, so the plasma would be destroyed. On the other side of the pane the plasma, including any stabilizing structures, would have to be recreated in a way which allows the BL object to form again, but there is no reasonable way to achieve this since the energy in the plasma will already have been spent.

Microwave Bubble Models

As we have seen, the observations of BL objects passing through glass windows eliminate from consideration all theories based on matter in some form or another. Only one class of models remains: those based on electromagnetic radiation. We are all familiar with radio waves and microwaves (from WLAN, for example) entering through windows, so if the energy of a BL object is stored in an electromagnetic radiation field, it would be able to pass. The problem is that microwaves propagate with the speed of light, so in order to keep them in one place, one needs reflectors on all sides. In the case of a microwave oven, the metal walls act as mirrors for the microwaves, but for BL objects, there are obviously no metal walls available. However, there is another possibility.

For a reflector of electromagnetic radiation, one needs free electrons that can be moved by the incoming waves. In the case of BL, one must replace the metal, where the electrons can move freely as in an electron gas, by a plasma in which unbound electrons are also available. As discussed before, the electrons in the upper atmosphere, the ionosphere, reflect radio waves below a frequency of about 30 MHz, so that could serve as a model for the walls of a microwave bubble BL. Unfortunately, trapping microwaves in a plasma shell is harder than trapping radio waves: one needs a much higher electron density. Another problem is the fact that the electrons must basically be in a collision-free environment, otherwise they could not follow the incoming radiation unhindered. In other words, they must be in a vacuum. This is obviously no problem in the thin upper atmosphere, but on ground level where the BL objects are observed, some mechanism is required which pumps the air out of the BL volume to make it a vacuum.

Despite these obvious difficulties, several theoretical models for BL have been proposed which postulate a well-conducting plasma shell around the microwave core (Fig. 11.1). The plasma shell is then the only part of the BL object that is visible, because the microwaves are of course invisible. The energy storage is due to the microwaves and the plasma shell provides the stabilization. Some of these models are Dawson and Jones (1969), Jennison (1990), Endean (1997), Zheng (1990), Handel and Leitner (1994), and recently Wu (2016).

As indicated above, the requirements placed on the reflecting walls are quite severe. In order to achieve the BL lifetimes of several seconds, a resonator of excellent quality is required with extremely low losses due to its resistivity. A thorough analysis of this model with emphasis on the resonator requirements is given by K.D. Stephan (2016). It turns out that the quality of the cavity must be extremely high, it has to be comparable to the quality of the very carefully manufactured, superconducting cavities used in particle accelerators like the one at CERN, Geneva. Here niobium cavities are employed, cooled down to 4 K with liquid helium to render them supercon-

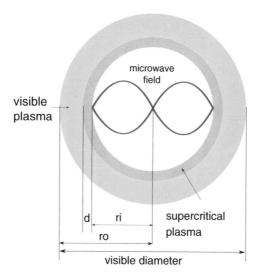

Fig. 11.1 Microwave bubble model of ball lightning. The supercritical plasma is the reflector of the microwaves and is surrounded by the visible plasma shell. Drawing by the author

ductive. The key to achieving this goal is a so-called supercritical plasma which is free from neutral atoms or molecules. The thickness of the reflecting plasma shell is given by Stephan (2016) as being in the range of 1–10 cm. The vacuum is created as follows: the air in the cavity becomes ionized by the intense fields and both electrons and ions are expelled from the high field region by the radiation pressure of the microwave field, so the inner part of the BL object would be at least a partial vacuum. Air trying to enter from the outside would be quickly ionized in the surrounding plasma layers and kept away from the core. The visible surface of the ball is then due to the outer plasma shell, where the air molecules are excited by the electrons and radiate light. When the energy stored in the microwave field becomes too low to sustain the reflecting plasma wall and the outward going radiation pressure, the shielding fails, and either the remaining energy is quickly radiated away as a burst of microwaves or it ionizes the incoming air almost explosively. In either case, if the cavity is rapidly filled with air, the implosion will be heard as a bang.

How can such an object pass through a window? The microwaves pass without problems, but the plasma shell cannot penetrate the glass, it is stopped on one side and must be recreated on the opposite side of the pane. It may be that charges attracted to the opposite surface start a surface discharge and that the passing fringes of the microwave field then use the free electrons of the discharge to reproduce the required shell, but even this scenario is not without

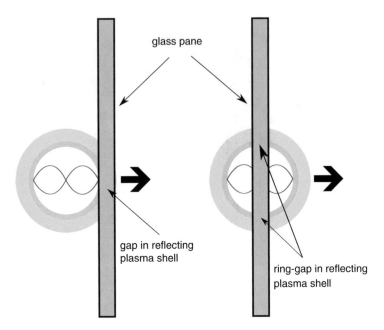

Fig. 11.2 Thought experiment: microwave bubble ball lightning passing through a glass pane. Drawing by the author

a fatal flaw. Even if the reproduction of the reflecting shell is working well, a small circular ring—the part in the glass—will always be completely without reflector. Assuming half of the ball has passed the window pane, there will be a half-sphere of plasma inside and a disconnected half sphere outside the glass, but at the position of the pane there will be a circular ring with no microwave reflecting properties at all (Fig. 11.2), because no plasma can exist inside the glass itself. This ring can work as an antenna for the microwaves which will quickly be radiated away. The thickness of the ring corresponds to the thickness of the glass, which is typically four millimeters. Double-pane windows will create a thicker ring depending on the type of gas filling between the panes, or a double ring with a plasma in-between.

We see that even for the plasma bubble models, passage through windows is impossible.

A model similar to the microwave bubble models has been proposed by Torchigin and Torchigin (2004), based on a spherical wave guide which reflects light while keeping it concentrated. He also explains how these objects could traverse a single glass pane and move through holes smaller than its original diameter. Unfortunately for this model, a BL object was observed at Neuruppin to pass through a curtain which subsequently showed signs of burning. A light bubble would not be able to pass through a curtain. This theory also has problems with a non-ideal atmosphere containing clouds, rain, or dust which would absorb the light.

Electromagnetic Interference Models

What is needed is a model where the main functions, the energy storage, stability, and localization, are all provided by an electromagnetic field. This sounds impossible, but it turns out that there is one model theory with all these properties. The first model proposed roughly along this line of thought was by the Nobel prize laureate P. Kapitza: he proposed that an interference of radio waves could lead to an electric breakdown of air and that the resulting plasma ball is what is visible as BL. The essential ingredient for this model is a strong electromagnetic wave emitted from lightning, then reflected, for example by the ground. The counter-propagating waves will create a standing wave pattern, where the strong field at some points, the nodes, ionizes the air. This model has been very influential in BL research because it promised to resolve the BL energy problem: the energy is continuously fed into the plasma ball from outside. Unfortunately, radio waves of the required strength have never been observed. The strength needed is in fact so high that other effects like damage to electronic equipment or even bodily injury would be more likely.

The Only Theory Remaining

Kapitsa's model of interfering electromagnetic waves obviously does not work, but the basic idea has influenced several scientists who have explored other possibilities involving electromagnetic waves, and this has led to a very intriguing model. In order to explain it, we must go back to the year 1865. This was the year, that Scottish scientist James Clerk Maxwell published the equations that finally provided a unified theory of electricity and magnetism. These four equations are one of the cornerstones of modern physics, constituting the basis for optics and electromagnetism. A surprising prediction of these equations was that, even in vacuum where there are no electrical charges, there are solutions of the equations that describe electromagnetic waves propagating at the speed of light. In the period from 1886 to 1888, Heinrich Hertz succeeded in proving the existence of radio waves by performing a series of experiments with a transmitter and a receiver driven by spark gaps. This paved the way to radio, television, radar, WLAN, and all the myriad of applications of EM waves we are now familiar with. In addition, Maxwell's theory of electromagnetism provided the first example of a unification: it combined both electric and magnetic effects in one coherent theory, providing the blueprint for the program which is still followed today in physics, where the electromagnetic force and the weak force have been unified as part of the standard model of particle physics. The unification of these with the strong nuclear force and gravitation has not yet been achieved and is one of the big challenges in physics today.

The Maxwell equations in vacuum leave two possibilities for electric field lines: either they can extend infinitely, or they form loops in a finite space. The first solution are the familiar electromagnetic waves, but the second possibility has curious properties and it has not been much explored as yet. The first to propose such solutions as a model for BL was Arnhoff (1992), and this was then taken up in 2002 by Chubykalo and Espinoza (2002), Chubykalo et al. (2010) and more recently by Cameron (2018). Cameron calls these exact solutions of Maxwell's equations "unusual electromagnetic disturbances", because they are not electromagnetic waves propagating though space, but they have looped electric field lines of finite extent and a localized appearance in all three spatial dimensions. These objects behave in fact more like particles than waves. This is precisely what is needed for BL objects that can pass through windows: the energy is stored in the electromagnetic field, and the stability is also provided by the field configuration, so no external reflector is needed! In this model, the visible plasma shell of a BL would only be a sort of "decoration" of the central field configuration, but it would not be responsible for the stability of the BL object. How can one imagine such a field configuration? In his paper, Cameron (2018) shows several nice images of the electric field distribution of the configurations he studied. Figure 11.3 shows the electric field vectors of the electric ring configuration and Fig. 11.4 the electric field of a globule configuration. More complicated configurations can be constructed, for example by superimposing simple versions. This allows for "tangle" configurations (Fig. 11.5), knotted ones, straight or curved lines, interlocking triangles, and so on. In fact, a whole zoo of different shapes can be built, using simple configurations like Lego blocks to construct more complicated ones.[2] Some of these objects can be seen as superpositions of plane electromagnetic waves, but others are more complicated and cannot be envisaged in such a way. The last two objects, the electromagnetic globule and

---

[2] The Maxwell equations allow the combination of simpler solutions into more complicated ones by linear superposition.

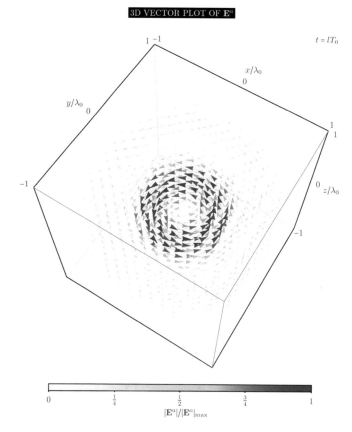

Fig. 11.3 Electromagnetic ring configuration, showing the field vectors of the electric field. During the next half-wave, the direction of the field will be reversed. From Cameron (2018), license: CC BY 3.0

the electromagnetic knot, have shapes resembling a spherical BL object. They do not move with the speed of light like "normal" electromagnetic waves, and can stay completely at rest or move with any speed less than the speed of light. If such an object is at rest, no energy is transported through space and the energy remains confined.

How can such a localized electromagnetic structure, which is invisible, produce the visible envelope? If the electric fields in the core of the structure are high enough, they can accelerate electrons to speeds at which they may ionize air molecules, creating an air plasma. The radiation pressure will push electrons and also ions out of the high field region toward the periphery of the object, where they accumulate and constitute the visible shell of the BL object. So far, the theoretical model has not considered this interaction of electromagnetic fields and the plasma.

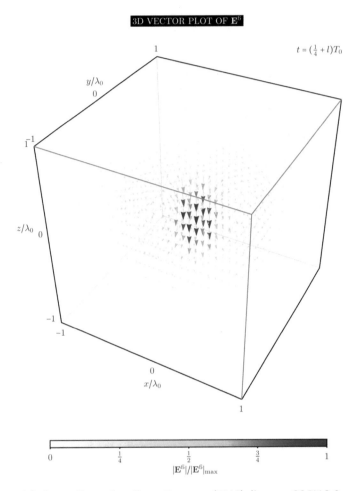

Fig. 11.4 Globule configuration. From Cameron (2018), license: CC BY 3.0

The appearance of the BL object will certainly depend on the specific structure of the electromagnetic object, and especially on its structure close to the surface. The various reports display a considerable range of optical appearances for BL objects, some being optically dense, while others are transparent. Some have a smooth surface, while others sparkle. All these different visual appearances may be produced by variations in the distribution of field lines inside the object.

In order to see whether these things really do match the physical structure of BL objects, we have to check thoroughly against the observed characteristics. We will see now how the model fares with respect to the list of essential features.

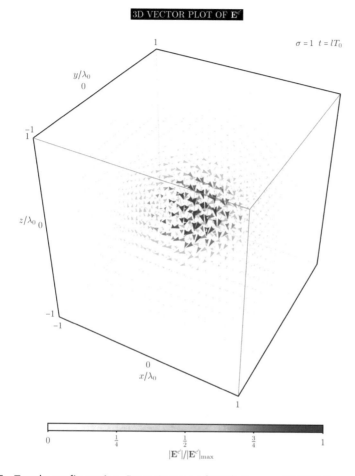

Fig. 11.5 Tangle configuration. From Cameron (2018), license: CC BY 3.0

The correlation with thunderstorms, especially with linear cloud-ground strokes (and preferably positive CG strokes). Creation at a distance from the lightning channel The EM structures are not electrical discharges but waves, so they are created differently: one needs a transmitter and an antenna. We shall see in the next chapter which structures occurring in thunderstorms have the potential to produce these special electromagnetic waves. As there are several potential solutions, we can tick this point off.

Their stability for seconds, up to minutes The long-term stability poses no problem for these structures; without energy loss due to the radiation of the plasma shell, they could basically exist forever. So we can tick this point off as well.

Their ability to store energy Electromagnetic waves can be very weak, but they can also be very strong. Radio signals from objects in outer space need very sensitive receivers just to record them, but on the other hand microwaves are used for heating food. The electromagnetic field can store energy up to very high levels, at least up to the levels that have been reliably observed for BL objects. Check this point as well.

Their stable spherical shape in more than 80% of cases, but also ellipsoids, toroids, or even irregular shapes One very interesting property of this BL model is the fact that the shapes of the objects and their structure are not restricted to spheres. Other shapes are also possible, which explains the rarer observations of objects with ellipsoidal, toroidal, tear-like, or even less symmetric shapes. It may also explain some BL objects where, through a transparent shell, some inner structures were visible, as in Case 30 described by Brand (Case 6 in appendix). Such an object may just have had a structure with several concentric shells of strong electromagnetic fields, and the visible plasma may have occurred in a certain region inside the object as well as outside.

Of course, there are other factors that play a role in determining the shape of these structures. Spheres have the smallest surface for a given volume, so this hints at a sort of surface tension trying to minimize the surface.

Check this point as well, but keep in mind that there are open questions concerning the prevalence of spherical shapes.

The color and structure of their visible surface: sometimes smooth, sometimes sparkling, some BL surfaces are almost transparent, others are opaque In this model, the plasma envelope of the EM structure is just a decoration, it does not contribute to the stability of the object. Trace impurities in the plasma will influence the color, as well as the amount of energy that the EM structure is able to give to the envelope. The precise structure of the electromagnetic object will also influence the surface structure, which is sometimes described as smooth, but also as sparkling or effervescent, or as consisting of constantly changing glowing discharges (Case 12). These discharges point to a strong electrical field parallel to the surface of the BL. In principle, this model can explain these features, but the precise details must first be analyzed. One has to keep in mind that, so far, the model is based on a mathematical description of the EM core, whereas the details of the plasma envelope and its interaction with the core have not been studied. Another property that has been described on several occasions is the splitting of BL into several smaller objects (Case 9). It is also conceivable that the field distributions within such an electromagnetic structure could rearrange themselves in such a way as to create several smaller objects with a lower net energy, but this is another feature that has not yet been studied. Basically, we can tick this point off as well, but keeping in mind that there is still a lot of work to be done to model the interaction between the core and shell, the creation of the shell, and the structural stability of the EM core.

Motion in open space, sometimes as if steered by some controlling agent The motion of the BL object will be controlled by the sum of all the forces acting on the structure. In this model, the plasma in the shell is not so dense that all electric field lines in the core structure will terminate at charges in the envelope. The major part of the electric field will be able to escape and interact with surrounding objects. From metal, the field will be reflected, producing a force that pushes it away; if the object has a partial vacuum inside, it will experience a buoyancy forcing it upward. The net effect depends on the nature and geometry of all surfaces within range of the escaping fields, and of course on the strength of the fields. We can tick this point off, but again, there remains a lot to be done to understand the forces that can act on these objects.

Their passage through glass window panes or other dielectric materials, and sometimes also through metal screens As already discussed in detail, the passage through dielectric materials like glass is no problem for an electromagnetic structure, but the recreation of a plasma envelope on the opposite side of the pane has to studied in more detail.

Metal screens, on the other hand, present a much more difficult obstacle for these structures. Metal screens are used in the doors of microwave ovens; they block all waves that have a wavelength longer than the size of the holes. Observations in which a BL object has passed through a metal screen report either no damage to the screen or the creation of a small hole where the metal was melted. The hole could be due to the local heating by the intense electromagnetic fields. Once the hole was there, it could have acted as a small antenna. There is a theorem, called Babinet's theorem, that states that holes in metals can act as antennas exactly like the parts that have been cut out in making the hole. So, if the wavelength of the EM structure is small enough, it could have passed through the hole. This behavior is like the disappearance of BL objects through keyholes, also reported in several cases. Another possibility is that the EM structure may be more extended than the metal screen. It is very likely that the screens were fixed in wooden frames, making them mechanically stable but electrically floating, since they are not grounded. Such a floating piece of metal could not block the EM structure like a full Faraday cage, because there is no complete metal enclosure. It appears possible that the EM fields, extending much further out than the screen, could just have interacted with it in a way that enabled an apparent passage. This is the most serious problem faced by this model, but it is not an insurmountable one. We can tick this point off for the passage through windows, but the passage through metal screens must be analyzed more thoroughly.

Their existence in closed spaces like rooms or aircraft As mentioned above, BL objects seen inside all-metal aircraft present an especially hard problem for almost all theories. Stenhoff lists several such reports, and one of my acquaintances has experienced a similar event (see the introduction to the book). The motion of the BL object described in these reports is almost identical: the object starts at the front of the plane, moving down aisle at the center of the aircraft towards the rear, where it disappears.

The skin of a modern aircraft is conductive, providing a Faraday cage which protects the interior from the effects of lightning strikes. However, it is always somewhat leaky because of the many holes like windows and cable feedthroughs. It also strongly attenuates radio waves coming from outside. The electromagnetic waves of these localized structures could only enter through the holes in the metal skin, which can act like inverse antennas, like key holes, as mentioned above. There is one account of a BL object becoming visible directly behind the front window, so it may have been formed there (Lowke et al. 2012). The motion down the center of the plane could be due to the steering effect of the electric field extending beyond the plasma envelope.

In the next chapter, we will come back to the question of how these structures could be formed inside aircraft.

We cannot yet tick this point off. This is clearly the most puzzling aspect of BL observations and clearly requires more work in order to understand it.

Their explosive or quiet modes of decay The plasma envelope radiates light away continuously so there will come a point in time when the EM core runs out of energy. This is at least true if no energy is supplied from the outside. The BL object could then just disappear without trace. If the energy stored and spent was high, there may remain some reactive gases which would give the characteristic odor sometimes reported. If the BL object had a partial vacuum in the core, there could be some sound: if the breakdown of the structure happened very quickly, the implosion would create a sound that could not easily be distinguished from an explosion.

The breakdown of the localized structure could also produce normal, plane electromagnetic waves. There is at least one report where the decay is described as explosive, bringing a wave of heat (O'Reilly, Case 3), which could be due to a burst of microwaves.

Again, we can tick this point off.

All in all, the model based on localized electromagnetic structures fits all the required characteristics, some very well, like stability and energy storage, others less well because the model has not been completely developed. Going back to the puzzle analogy, we can use almost all the pieces provided by our observations to construct a single image; we do not need to consider other theoretical models to account for certain special features.

Nevertheless, I am sure that the discussion of the theoretical BL models in this chapter will provoke a lot of criticism. Many people will find both the requirement of only one model that fits all observations and the three key requirements—creation away from the lightning channel, passage through window panes, and creation and existence in aircraft—far too restrictive. My justification is that, in view of the almost stagnant state of research in the BL field, a new and more radical approach must be attempted, and in particular, one that is rooted in the available observations. The outcome justifies the procedure: we are now left with one model, solidly based on existing physics, which fits more of the observations than any other model, and which above all fits them better.

This quite surprising result gives a clear message about how we should proceed: we need to find out how to make these objects in the laboratory. Unless we can produce them in a repeatable way, we cannot reliably test this hypothesis.

Experiments will be discussed in the next chapter, where we first analyze the conditions under which such objects might be produced.

Summary

- BL theories should be able to explain the central problems of BL modeling: energy storage, duration, shape, motion, passage through non-conductors, existence in aircraft, etc.
- The main groups of models are: chemical reactions (combustion), electrical discharges, plasma structures, microwave bubbles, and unusual electromagnetic structures.
- Almost all proposed BL models can be eliminated by checking against three well established observational facts: creation far away from lightning channels, passage through glass windows, and creation and existence in aircraft.
- All but one fail these criteria: even models proposing a high-frequency resonator filled with EM radiation are inconsistent with these observations.
- Only one theoretical model remains: special solutions of Maxwell's equations with looped electromagnetic fields, having a particle-like character (Cameron, 2018).

- These models can explain other unusual properties of BL as well, including their motion and the variability in their appearance.
- The relation between the electromagnetic core and the visible plasma envelope has not yet been worked out.
- BL objects passing through metal wire screens looks possible, but will require further investigation.
- BL creation in modern aircraft remains possible but will require further insight into the nature of these structures.

12

BL Experiments

The final goal of BL research is the repeatable production of such objects in the laboratory and their controlled study, something that proves impossible "in the wild" because of their unpredictable occurrence. Unfortunately, the number of experiments dedicated to BL production is much smaller than the number of theories about them. Making theories of course is easier and cheaper than doing experiments, since the latter requires a laboratory and an experimental setup (and maybe a considerable amount of money), whereas theories can in principle be made in your study with paper and pencil alone.

In this chapter we will first look at how to make the localized electromagnetic structures which we singled out in the previous chapter as the most promising theoretical model for BL. The fact that we have a clear idea of what we are looking for gives us enormous leverage for the definition and analysis of situations where BL might arise, so we will check how the naturally occurring conditions in thunderstorms could produce such objects.

Then we will look at the experiments, which fall broadly into two categories: planned ones, where the researchers intended to make BL objects and accidental ones, where objects resembling BL were produced by equipment not intended for that purpose. We will first take a look at the accidental ones, because it is particularly interesting that there may even have been real BL among the luminous objects produced.

© Springer Nature Switzerland AG 2019
H. Boerner, *Ball Lightning*, https://doi.org/10.1007/978-3-030-20783-0_12

How to Make Localized Electromagnetic Structures

The localized electromagnetic structures we encountered in the last chapter are in principle electromagnetic waves, but they interfering with one another in such a way that they create confined structures that do not extend with the speed of light. Electromagnetic waves are normally produced by a transmitter coupled to an antenna. Figure 12.1 is a schematic of such a well-known setup.

The oscillator produces a sine wave of the desired frequency, which is then amplified, and the signal is sent to the antenna, which radiates it away. What is responsible for the creation of the radiation are the free electrons in the metal of the antenna. They are driven by the electric field set up in the antenna structure by the transmitter, and they are continually accelerated, going up and down. This acceleration is essential because, if we had a constant current, a magnetic field would be created but no electromagnetic wave. The structure of the antenna determines the radiation pattern of the waves; here it is a vertical dipole which radiates horizontally in all directions. Far away from the radiating dipole, the wave fronts can be considered flat, so we then have plane electromagnetic waves moving forward at the speed of light. Figure 12.2 shows how the electric and the magnetic field of such a wave vary with time.

We can now define some of the essential ingredients needed to create the electromagnetic structures:

- free electrons than can oscillate,
- a configuration of the electrons of a suitable shape to act as an antenna,
- energy to drive the oscillations of the electrons and create the electromagnetic field which then contains a part of the energy put into the structure.

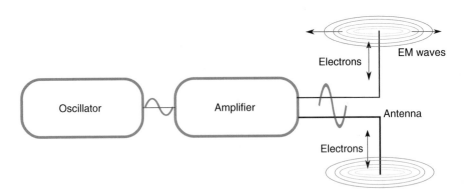

Fig. 12.1 Classical transmitter-antenna configuration. Drawing by the author

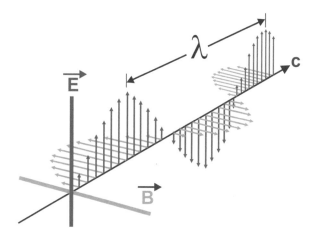

Fig. 12.2 Plane electromagnetic wave. Image by helder100 from Pixabay

As discussed above, there are several observations which report the creation of BL objects "out of thin air". To create them in a given place, we need free electrons in the ambient air, since obviously no metal antenna was available. Note that an antenna shaped like a dipole would not do the job, because it would just radiate the energy away, so a rather special shape is needed. Cameron (2018) gives an example of how an antenna for an EM object might look (see Fig. 12.3). We see that the antenna is a closed structure that radiates inwards to produce the desired field configuration in the center, so an electron cloud of roughly spherical shape, probably hollow, is needed.

We need free electrons because of their high mobility, but electrons in air normally have only a very short lifetime. At normal pressure, it takes only about 100 ns before they are caught by oxygen molecules (Boissonnat et al., 2016) and water molecules, which tend to attach electrons in collisions, thus almost immobilizing them. Even worse, the negative molecules tend to collect water molecules around them, making them much heavier still. In fact, the electrons start condensation nuclei which seed the formation of tiny water droplets. This can be seen very well in the so-called cloud chambers, where the electrons are kicked out of molecules by passing charged atomic particles, producing tiny water droplets in the saturated atmosphere which then mark the track of the particle by a fine trail.

It is really not easy to create a cloud of free electrons. As discussed above, the positive coronas, where the discharge is started from a positively charged conductor, produce positive ions, while electrons are collected by the positive conductor. Only negative coronas can produce large amounts of electrons, but these move away from the negatively charged conductor into regions of

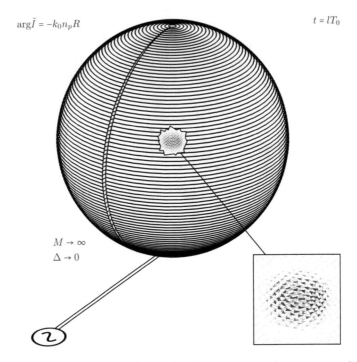

$\arg \tilde{I} = -k_0 n_p R$

$t = lT_0$

$M \to \infty$

$\Delta \to 0$

Fig. 12.3 Antenna for creation of a localized electromagnetic structure at its center. From Cameron (2018), license: CC BY 3.0

lower electric field where they are scavenged by oxygen molecules. In order to create a large cloud of free electrons, the electric field must be so high that the attachment process becomes unlikely. Recent data show (Boissonnat et al., 2016) that a field strength above 50 kV/m is required to produce an appreciable fraction of free electrons in the negative space charge region.

In order to envisage how this process could evolve, we take a look at a situation which—as we have seen—yields a high probability for BL generation: a positive CG lightning stroke. In this scenario, a positive leader will descend and create an electric field at ground level that rises slowly at first, but then intensifies rapidly until a maximum is reached when an upward going negative leader is created from the ground. This upward going leader is practically at ground potential, so the electric field around it is already reduced. When the downward moving leader contacts the upward moving connecting leader, the return stroke starts and the field breaks down almost immediately. The negative space charge could be created rather early during the initial breakdown in the clouds, but only in the final phase of the approach will the field be high enough to make free electrons and accelerate them sufficiently for them to move upwards and form a cloud that could act as an antenna.

Where could the electrons come from in the first place? Since there are few free electrons in the air, they must be supplied by a slightly conducting surface, ideally a metal surface. Metal structures will be able to supply more current and they can also produce secondary electrons by the photoelectric effect, whereby ultraviolet photons radiated from excited molecules kick out more electrons from nearby metal surfaces. In this respect, Case 12 is particularly interesting: here a BL object was created above a rather large circular brass table. Moreover, the BL had an unusually large diameter of about 1.5 m, which corresponded more or less to the diameter of the table. This suggests that the metal table started the electron cloud already roughly in the shape needed for the formation of the antenna which then created the electromagnetic structure of the BL. There are other cases where the BL was created near to curved metal surfaces, like a door knob or a stove pipe. In Russian BL collections, the BL is often created near metal lines or wires. So, the correlation between a metal surface able to start the electron cloud and BL is clear, but there are many cases where there was no metal object underneath the position of the BL, as with witness reports 16 and 20 of the Neuruppin event (Case 1). Here it is less clear how a sizable space charge could have been created.

Another problem is the time interval in which the electron cloud could have been formed. In Neuruppin and in Case 12, the BL objects were formed about 1–2 m above the ground, so the electrons must have traveled this distance during the time when the electric field was high. The down-coming positive leader has a typical speed of about 400 km/s, so the last kilometer down to ground will have taken roughly 2.5 ms.

To calculate the height that negative ions and electrons could have attained when starting from the ground under the influence of an intense electric field, we take two values as examples: the value where the isolating properties of air finally fail and a more realistic figure of 1/10 of the breakdown field, i.e., 3 MV/m and 300 kV/m. Negative ions moving under the influence of the lower field value will cover 6 cm in 1 ms and about 60 cm for the breakdown field value, so they are rather slow. Electrons, on the other hand, can move much faster: they will move about 23 and 140 m, respectively, in 1 ms. The conclusion is that electrons can easily move to the height above ground required for the indoor observations of BL objects in Neuruppin, but that for higher field values, they are probably too fast, i.e., they would move away too rapidly. So there will be a "sweet spot" around the strike point of a positive CG lightning where the electric field is just right for the formation of a negative space charge cloud: closer to the lightning channel the field is too high, so electrons move too fast and are also likely to create negative streamers; but further away the field cannot raise the electrons to a suitable position during the approach of the down-coming positive leader.

Of course, the structure of the space charge region will not only contain fast electrons. During their acceleration, they will collide with molecules in the air, exciting them, breaking them apart, or ionizing them, creating secondary electrons and positive ions in their wake. The positive ions will move in the opposite direction from the negative charge, building an electric field between them. The detailed structure of this region will therefore be quite complicated, but since all the basic processes that occur in this scenario are known, one could in principle simulate it on a computer. However, this has not been done so far. One the other hand, simulation of the development of streamers is an active field of research, since streamers are the fundamental building blocks of electric discharges, and understanding them helps to shed light on several processes which occur during the development of lightning discharges. Normally, especially for strong positive lightning, we would also expect negative streamers to develop instead of the space charge region, even if they require a higher electric field than their positive counterparts. In Neuruppin, this appears to have been the case closer to the hit point of the lightning, since at least one of the witnesses described "huge blue bundles stretching towards the sky" (Case 1, Witness 5). The cloud of free electrons that can act as an oscillator/antenna combination to produce the electromagnetic structures can only be formed under conditions which inhibit the production of streamers. In particular, sharp, pointed objects, which produce an intense electric field at the tip, must be avoided, because they are good candidates for starting a streamer. In Case 5, strong corona production was reported all over the village, while a BL object formed over a water surface which was rather smooth. There are other cases where the BL appeared over a rather smooth surface, as in Case 13 in Australia, where the object appeared over the surface of a road, or in Case 12 where the BL appeared over the brass table.

We now have a suitable space charge cloud of free electrons, but how can this be converted into a BL object? In a case like Neuruppin, where the objects were created far away from the lightning channel, another source of energy besides the electric field will be required, because the energy density of the electric field is rather low. For a field strength of 3 MV/m, the breakdown field of air, the energy density is about 40 J/m^3. Only rather weak BL objects could be created by such a source of energy, so are there other sources that could have been tapped? In fact, there is another possibility. Lightning of the strength observed in Neuruppin tends to emit a very strong pulse of wideband electromagnetic radiation, a so-called electromagnetic pulse or EMP during the return stroke. In principle, every lightning strike emits such a pulse, but with strong lightning, the EMP creates "elves" in the upper atmosphere, as explained earlier. The EMP created by the return strokes propagates upwards as a spherical wavefront and intersects the ionosphere in a circle. There, it excites the free electrons which heat and excite the gas molecules. The luminous ring produced by this heating appears to expand faster than the speed of light. The EMP of very strong lightning, above 250 kA maximum current, was even observed to have a long-lasting influence on the ionosphere above the position of the lightning (Haldoupis et al., 2013). Can the negative space charge cloud, which is much closer to the source of the EMP than the electrons in the upper atmosphere, be excited in a similar way? It may be possible that, at the position where the free electrons are available, a plasma is formed that initiates the transformation of the propagating electromagnetic waves into a localized electromagnetic object. In this model, the electrons in the space charge cloud are essentially the receiver of the incoming EMP. How they could transform the energy from the EMP into the energy of the localized electromagnetic object forming the core of the nascent BL is of course an open question.

Besides positive CG lightning there are other situations where BL is created, but with a lower probability. In particular, negative CG lightning, sometimes of rather low intensity, can also produce BL. In a recent case, a rather weak negative CG lightning stroke of only 15 kA created a BL object which was seen to move through windows (Case 15). Here the region under the approaching stepped leader will create a positive corona, and finally, positive upward-going streamers that grow into connecting leader. Production of a negative space charge under such conditions would not be possible, so we need to look for another scenario. Such a situation arises with the return stroke: now the huge amount of negative charge stored in the corona sheath of the lightning channels flows down to earth, biasing it negatively with respect to surrounding objects. Negative corona and flash-overs can occur, and when the lightning or a flash-over strikes a metal object like an open-air telephone or power line, negative discharges can occur far away from the strike point of the lightning. It is therefore possible that at parts of these lines with a suitable geometry, a negative space charge could be created with the shape required for an EM structure. In this respect two observations are important: the appearance of BL is often reported from electrical sockets (Grigor'ev et al., 1992) or, especially in the US, from the microphones of old-fashioned crank telephones (Cases 21 and 23). In these cases, a high voltage pulse from a close lightning stroke may have created a discharge at the socket or the microphone which was the source of a space charge cloud and then the BL object. Figure 12.4 shows one of these telephones. The microphones had

Fig. 12.4 Old telephone with a funnel-shaped mouthpiece. From: Tomasz Sienicki (https://commons.wikimedia.org/wiki/File:Telefon_VHM_ubt.jpeg), "TelefonVHM ubt", CC BY 2.5

a funnel-like mouthpiece that may have assisted in creating a spherically shaped electron cloud which would subsequently have emanated from the funnel.

What about BL creation close to the channel of a lightning stroke? There are at least two cases where such a process has been documented photographically, by the cameras of the meteorite network and by Ern Mainka (Case 2). As described above, the lightning channel produces a strong corona in a region around the conductive core, the so-called corona sheath. For a negative lightning stroke, this corona sheath will of course contain a strong negative space charge of free electrons that is drained towards Earth in the return stroke. So, space charge and an ample supply of energy would be available in this case, but there is a problem: the corona sheath will be formed by a multitude of thin streamers emanating from the conductive lightning channel, and not an electron cloud with the correct shape. Basically, the lightning channel and its corona sheath will form a cylindrical capacitor. When the electrons are not arranged in, say, a rather spherical shape, they will not make a radiation field that is suitable to create localized electromagnetic objects. It is clear from the low creation probability that will this will only be possible in rare cases, but in those cases BL objects of very high energy content can be created.

Here we can take a quick look at the problem of BL creation inside aircraft. We have established that a cloud of free electrons and energy input is needed for BL creation, so could these exist inside an aircraft? The electrons cannot have been produced by an electric field as they would in the case of a positive lightning stroke approaching the ground, because the Faraday cage provided by the all-metal aircraft prevents this, but during ascent and descent, aircraft pass through the level in the atmosphere where cosmic ray air showers are most pronounced. Again, the aluminum skin of the airplane will shield the interior to some extent, but there will certainly be a burst of electrons and ions when the airplane suffers a direct hit. In modern aircraft, the cabin is pressurized, but the pressure is lower than at ground level in order to decrease the force on the cabin walls. Normally, the pressure is about 30% lower than at sea level. This lower pressure increases the mean free path for electrons by this amount, and it decreases the probability of attachment to oxygen molecules even more, since this attachment involves the collision of three particles, the electron and two gas molecules, in order to satisfy energy and momentum conservation. It is therefore easier for electrons in the atmosphere inside an aircraft to remain free, and they can also gain more energy between collisions. When an aircraft suffers a direct hit by lightning, strong currents will flow through its skin between the points where positive and negative leaders are attached. Such current pulses can excite electromagnetic modes in the structure of the aircraft, which acts as a resonance cavity with its metal walls. These electromagnetic waves can act on the electron cloud created by the cosmic ray shower. In principle, the ingredients for BL creation may be here, but the details involved in the production of such objects under these circumstances are of course not yet clear.

We have seen now that localized electromagnetic objects could perhaps be created by positive and negative CG lightning above the ground and close to the lightning channel in the air. Thus, the hypothesis that there is only one type of BL still holds, but for its creation we have to modify the original assumption somewhat: the free electron cloud acting as an oscillator/antenna combination can arise in all these cases, but the exact process involved may be different according to the type of lightning under consideration.

The big question is now: how could one duplicate such a process in the laboratory. And why has this not been possible up to now? There many high voltage (HV) laboratories around, so why has BL production never been reported from any of these labs? This negative result is sometimes used by skeptics as an argument against the very existence of BL, but is it really true that the conditions in thunderstorms and in the laboratories are comparable?

To try to understand this, let us look at some of the experimental setups used in such labs.

In order to simulate the influence of lightning strikes on planes or electronic equipment, HV pulse generators are used. In most cases, a so-called Marx generator provides the voltage pulse (see Fig. 12.5). Such a generator is basically a bank of HV capacitors which are charged via a network of resistors to a DC HV power supply. During charging the capacitors are biased in parallel, so they are all charged to the same voltage. Then a series of spark gaps is triggered, which very rapidly couple the capacitors into a serial configuration, adding up the individual voltages to produce a multiple of the voltage at the output. Normally, a discharge then starts between the HV terminal and the object to be tested. The spark quickly discharges the capacitors, and the resistor network of the generator also equalizes the voltages, so the HV breaks down quickly.

A typical pulse shape used in such tests is shown in Fig. 12.6. The rise time is of the order of 1 μs and the decay time is about 50 times longer.

Obviously, this pulse shape only simulates the return stroke, but not the rise of the electric field produced by an approaching leader, which is of the order of milliseconds, about a thousand times slower. Other HV pulse generators can be used, but they all have a very rapid rise time which is unsuitable to reproduce the effects of the leader approach. The very short rise time does not allow enough time even for the fast electrons to form a space charge cloud of the required size and shape. Another factor detrimental to the production of a negative space charge cloud is the fact that the HV test equipment is usually designed to produce a linear discharge—first the streamer, then the leader, and finally the spark—between the HV terminal and the objects. Producing a negative space charge cloud by suppressing the spark, so that the cloud can act as an oscillator/antenna combination, is just not the aim of such a test setup.

Other situations or devices are also characterized by the very rapid rise of the voltage pulse. These include the sinister cases of nuclear explosions in the upper atmosphere, and so-called electromagnetic pulse weapons simulating the EMP of such explosions.

Nuclear explosions high above in the ionosphere create an enormous wideband pulse of electromagnetic radiation. The gamma rays produced by the explosion kick electrons out from the air molecules beneath the explosion site and push them away from the positive ions. This process creates a huge dipole which bathes a large area of the Earth's surface in its radiation. Such a nuclear EMP can destroy or damage electric and electronic equipment because it couples HV pulses into the devices. This pulse is far stronger than the EMP from

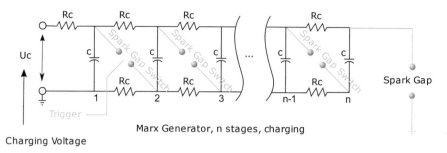

Fig. 12.5 Schematic of a Marx generator. Adapted by the author, from Wikimedia commons, public domain

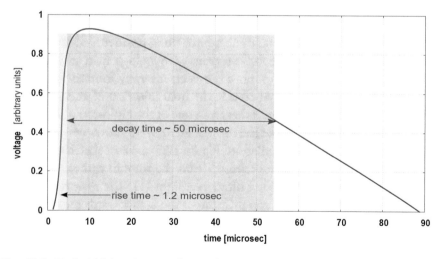

Fig. 12.6 Typical high voltage pulse used to simulate lightning in a laboratory. The rise time is about 1.2 μs (from 10% to 90% of the peak value), while the decay time (to 50% of the peak value) is about 50 μs. Drawing by the author

lightning. Since modern warfare relies increasingly on electronic equipment which is very susceptible to such HV pulses, all these devices must be adequately shielded. Despite the enormous strength of such EMPs, there is no report whatsoever that they have ever created any BL objects. There has been a considerable amount of research on devices which produce similar EMPs on a smaller scale to destroy military electronic equipment. Usually, these EMP weapons consist of a HV generator that produces a strong HV pulse. The pulse is then fed to a microwave generator that converts it into a burst of electromagnetic radiation in the GHz range. This radiation is concentrated on the target it is supposed to damage. Most of the research in this area is of course classified, but still, one would expect to have heard rumors if in the course of testing such devices, BL-like objects had been produced, and indeed no such reports have ever surfaced. If we compare the action of these electromagnetic pulses to the list of requirements for the production of localized electromagnetic objects, one problem is immediately clear: these pulses will not be able to create a cloud of electrons in air that can act as an antenna. They rise much too fast and, of course, they do not produce the quasi-static electric field that could move the electrons into a favorable position and shape. The combination of such an electric field and the subsequent EMP from the return stroke appears to be one feature exclusive to positive CG lightning.

This brings us to the last topic of this paragraph: can we define an experimental setup that could mimic the conditions for BL production existing in a thunderstorm? For the situation of an approaching positive leader, one needs a HV device that can produce free electrons with its high field and that has a pulse length that is sufficient to move the electrons away from the cathode so that they can form the required cloud. The final injection of energy into the electron cloud could then come either from one of the rapid-rise pulse generators or from the EMP of one of the devices mentioned above, which produce a broad-band pulse of GHz radiation. Figure 12.7 shows a sketch of the proposed setup. First, electrons are released from the smooth cathode (avoiding streamer formation) by a pulse of ultraviolet light (1). The electrons become attached to oxygen molecules and drift under the field towards the anode (2). After a suitable delay, a short HV pulse is applied which creates free electrons that produce an avalanche. At the end of the HV pulse, an EMP is applied (3). The free electrons are accelerated under the combined influence of the quasi-static field in the space charge and the field of the EMP, creating the BL object (4).

In the case of negative CG lightning striking metal objects, the setup may be simpler. One would need a HV pulse generator coupled to a suitably shaped cathode where the electron cloud is produced, but where the start of a

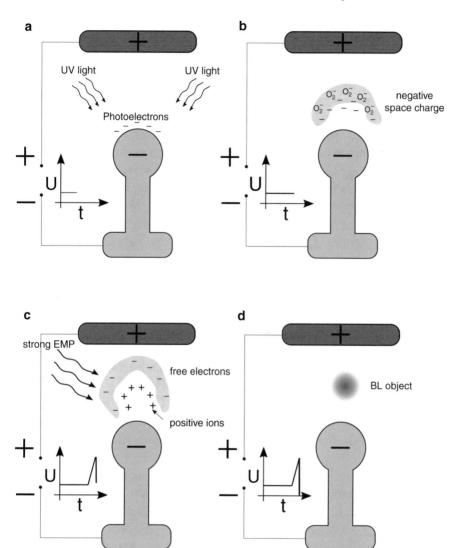

Fig. 12.7   Possible experimental setup to create BL objects. Drawing by the author

streamer is avoided. Once again, the pulse length has to be sufficient for the electrons to move an appreciable distance to form the antenna shape. The final injection of energy could come from a last rapid rise in voltage and an external electromagnetic pulse.

The reader may wonder whether this is really a reasonable recipe for such an experimental setup? This is just my educated guess[1] of how one could begin to design an experiment, but, of course, many important details will still be missing. One problem may be that these experiments do not produce a BL object every time they are fired. The space charge cloud may be different each time the experiment is performed and it may only work in a small number of cases, so such an experiment would have to be repeated many times.

Another parameter that could be changed in such experiments is the composition of the gas in which the experiment is performed. Since oxygen strongly suppresses the existence of free electrons in air, it may be a good idea to work in an oxygen free gas, such as nitrogen, argon, or a mixture of the two.

Armed with the list of requisites for BL production, we are now ready to look at the experiments, accidental or deliberate, that have been performed to make BL objects.

Accidental BL Experiments

The first of these unintentional experiments was probably performed by a researcher called Richmann in 1753. Richmann wanted to repeat Franklin's experiment with the kite to see for himself that lightning was electrical in nature. He placed a sort of lightning rod in a building in St. Petersburg, Russia, but he did not ground it, see Fig. 12.8. The metal rod appears to end on a sort of stool. When a thunderstorm approached, Richmann hurried to his laboratory accompanied by an engraver from the Imperial Academy; there were no photographers at that time. The engraver named Sokolov said that "Richmann was a foot away from the iron rod, when a pale ball of fire, the size of a fist, came out of the rod without any contact whatsoever. It went straight to the forehead of the professor, who in that instant fell back without uttering a sound". In a contemporary report it was stated that his left shoe had burst open and that there was a blue mark on his foot.

It may well be that the unfortunate Richmann discovered one of the easier ways to make BL. When a stepped leader was coming down, the upper end of the iron rod will have sent out an upward connecting leader of positive polarity. Consequently, the lower end of the rod in the room, not being grounded, will have emitted a negative charge cloud. The return stroke will then have deposited more negative charge into the room, creating both a BL object and

---

[1] "Educated guess" is a euphemism used by scientists for a guess which the author considers to be very close to reality. However, this can sometimes be put down to the author overestimating his/her abilities.

Fig. 12.8 The fatal result of Richmann's experiment with an ungrounded lightning rod. Public domain

a normal discharge that killed the professor. Naturally, the outcome of the experiment was widely circulated in scientific circles, so it was never repeated. It was also realized that it was essential to ground a lightning rod properly, so there was never another case like this. It may be a relatively easy way to create BL objects from negative lightning, but for obvious reasons, working with such a setup should be left to experienced lightning researchers.

Another unintentional experiment is presented in Brand's book. In a power station in Norway, luminescent balls were observed on several occasions. Tests in which a generator was short-circuited sometimes produced luminescent spheres that could be photographed. Brand reproduces two of these photos in his book, but unfortunately the quality of the reproduction is so bad that almost nothing can be distinguished. The Norwegian engineer reported two other cases of luminescent spheres that were created when over-voltages occurred. Without further information, it is not possible to find out what was really produced there, nor whether it was even related to BL at all, but we can very cautiously attempt a check against the criteria given above. In all these cases, a voltage pulse due to a short circuit appears to have been involved, so in principle one could envisage that an electron cloud may have been produced. The fact that this always happened at a device where more than enough power was available could also point in the direction of BL objects. On the other hand, modern HV equipment used for power transmission does not appear to produce such fire balls. Maybe today power surges are better suppressed than in the early days of HV engineering.

Another curious case was reported in 1962. A Mrs. Eunice Overend sent a letter to the journal "New Scientist",[2] giving the following description:

"I was fiddling with a faulty hair-dryer and inadvertently touched the live wires from one terminal to the half-case of metal so that the motor spun for an instant as the insulation melted. A beautiful blue bubble, perhaps 2 in. across, appeared above it and then floated down. It vanished as it touched the cloth but left a hole like a cigarette."

There is a rumor that a scientist, who had read this description, acquired several faulty hair-driers at flea markets and tried to repeat this "experiment", but without success.

In his chapter on BL, Rakov and Uman (2003) describes one instance (Case 18) where a luminescent ball was created from a shortwave transmitter in an army truck. The operator produced an arc at the antenna output of the transmitter in order to light his cigarette (!). He goes on: "As I leaned back to the sitting position again, up from behind the transmitter floated that "shimmering fireball". Shimmering, pulsing, blue of its fire, it then floated right at us (18 inches (0.5 m) in diameter) (of a truth, to now I can't say for sure whether I had yet shut the transmitter off or not). .

In retrospect, I would have to say that the fireball had to originate at the same antenna tie-in as where I had lit the cigarette from. It floated up from that very side where the antenna post lead was connected."

The wavelength of this transmitter was in the short-wave range of about 60 m, so it was most likely too low for the direct creation of an EM structure only half a meter in diameter, but the electric arc may have produced a much higher frequency. This "emitter" must then have coupled into the space charge cloud produced by the arc and produced the BL-like sphere. It's a pity the two soldiers were too afraid to repeat their experiment and take some photos.

Another accidental BL experiment was performed in the early 1960s in the main scientific laboratory of the company Philips in Eindhoven in the Netherlands.[3] The scientists had quickly cobbled together an experimental setup to study a pulsed helium laser, and they used a copper coil which happened to be available. Since the resistance of the coil was too high for the test, they just cooled it in liquid nitrogen to lower its resistance, and then placed it

---

[2] New Scientist, 13 September, p 579 (available online at Google books).
[3] G. Dijkhuis, personal communication.

on a wooden stool. When the high voltage circuit feeding the helium discharge tube was pulsed, a sparkling ball of about 5 cm diameter would sometimes jump out of the open core of the coil. Once again, apart from the testimony of one of the engineers operating the experiment, no photos exist, and it is not clear whether the objects observed had anything in common with BL or were only a look-alike.

In his book (Sagan, 2004), Sagan[4] reports incidents that took place at the Hill Air Force Base in northern Utah, where solid-fuel rocket motors of Minuteman rockets were periodically X-rayed to check for cracks. The Air Force used a linear accelerator from Varian to produce hard X-rays that could penetrate the rocket motor. For nearly 1 year, a ball of glowing blue fire (usually on days of clear weather) would regularly drop out of a space adjacent to the accelerator's high voltage power supply, usually when there were arcing problems. The fireball would always glide silently to the floor, float around the room, and then rise back up to the power supply and vanish silently. When some of the fireballs damaged electrical equipment, the Air Force shut down the accelerator and had the high voltage power supply replaced. The new power supply used solid-state electronics, whereas the old one had had vacuum tubes. With the new power supply, no blue balls were produced anymore. Sagan claims to have reports on these incidents, but there are no photos or independent evidence.

Another report by Sagan is about a sputtering equipment which generated blue luminescent balls, coming out from the exhaust pipe of the vacuum pump. The report claims that these balls were generated frequently (full account in the appendix, Case 11).

The famous researcher Nikola Tesla, inventor of the so-called Tesla coils or Tesla transformer (among many other things) reported that, with his high-voltage coils, he managed to create glowing spheres. This caught the imagination of many people, and there have been several attempts to recreate these objects. Tesla left no report with details about the setup or a description of the objects, so it is almost impossible to validate his observation.

Accidental shorting of submarine batteries, where the electrodes of the circuit breakers were overloaded and partially evaporated, has also been known to create luminous balls of green color. Their diameter was about 10–15 cm and their lifetime up to 1 s. These fireballs were green due to the copper vapor from the electrodes (Stenhoff, 1999). Could these luminous objects have been BL? The short-circuit involved very high currents but no high voltage, so any

---

[4] "Paul Sagan" is a pseudonym of Paul Sniggier. His book is fun to read, but contains little that can help to solve the puzzle of BL: it mixes too much fiction with the facts.

free electrons would have been restricted to the hot plasma column created by the arc, and the formation of an antenna in the form of space charge would have been very unlikely. Much more likely would have been the production of a cloud consisting of hot vapor which dissipated rather quickly.

All the reports on the accidental experiments share this property: they do not offer enough information on how to reproduce the results, and nor are there any photos or videos of the luminescent objects produced. The exception is the case in Norway where photos do exist, but even Brand was unable to obtain any detailed information on the tests which produced these luminescent spheres. One thing common to several cases involving electronic equipment was that vacuum tubes were used, rather than solid-state devices.

Planned Experiments

Around 1990, Japanese researchers tried to put Kapitsa's ideas to the test. They created microwaves with a high power (5 kW) magnetron and let them interfere in a cavity in such a way as to generate standing waves (Ohtsuki and Ofuruton, 1991). The cavity was filled with air at normal pressure. The intense electric field of the microwave radiation led to a breakdown of the atmosphere, and plasma structures of several types, including ball-like objects, were produced. One type of plasma structure managed to burn a hole in the aluminum end cap of the cavity, and a plasma flame escaped. These plasma objects needed a continuous input of microwave power; only one type could be observed for a short while after the magnetron was switched off. One type of discharge moved through a ceramic plate placed in the cavity. These experiments were interesting enough, but even the Japanese researchers had to concede that microwaves of the power used in the experiments had never been observed in nature. In fact, microwave ovens work with much lower power levels, usually lower than 1 kW, so a naturally occurring microwave source at 5 kW would certainly fry plants, animals, and people and destroy electronic equipment.

Recently, there have been several experiments with discharges in water, creating so-called plasmoids. These plasmoids are superficially similar to BL objects.

In these experiments, a high-voltage discharge is created starting from a central electrode in a water filled vessel. The plasma consists of water, air, and parts of the electrode, forming a ball-like cloud of hot gas that quickly rises up (Fantz et al., 2015; Versteegh et al., 2008) (Fig. 12.9).

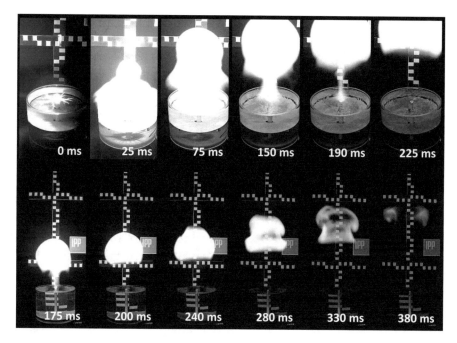

Fig. 12.9 Plasmoid created by discharge in water. With kind permission of U. Fantz, Max-Planck-Institut für Plasmaphysik, Augsburg

For about half a second the cloud displays a luminosity that is much greater than anyone once thought would be possible, but it does not have the structural stability of a BL object. It is also only composed of matter: hot air, excited or ionized atoms and molecules, and electrons, which means that it would be completely stopped by a glass plate or even a sheet of paper. These plasmoids have been very well studied (Fantz et al., 2015), and the conclusion of these researchers is (Versteegh et al., 2008): "The observed plasmoids arise from a hot expanding plasma with relatively high electron density of 10^{22}–10^{20} m^{-3} between $t = 0$ and $t = 75$ ms [...] Comparison with chemiluminescence spectra and the observed high vibrational excitation of hydroxide radicals suggest that dissociation products of water store the (chemical) energy enabling the autonomously radiating behavior. The plasmoid's cooler boundary layer consists of electric double layers that may contribute to the characteristic shape of the balls." It is obvious that these plasmoids are completely different from BL objects, but they could serve as models for the luminescent plasma shell of BL.

Plasmoid experiments are typical of attempts to simulate the natural lightning channel and its interaction with matter at the point where it connects with the Earth, assuming that evaporated and excited matter form the BL object. Similar experiments have used triggered lightning in order to come closer to a realistic simulation of this hypothetical BL production. As we have seen above, such models are falsified by observations that show that many BL objects are created far away from the lightning channel.

The reports concerning the production of luminous spheres from short-circuited submarine batteries prompted several other researchers to replicate this. One of them founded a company called Convectron in the Netherlands. The experiment involved a 200-ton assembly of submarine accumulators with a short-circuit capacity of 30 MW which reached a horrendous current of 150,000 A. Luminescent balls of 10 cm diameter and a lifetime of 1 s were produced and recorded on film. The aim of the company was to create self-stabilizing plasma structures that could act as fusion reactors for deuterium, but this goal was never achieved. As explained above, the experiments involved very high current but only moderately high voltage, so the creation of free electron clouds was impossible. The luminescent spheres were only pseudo-BL which had nothing to do with the real thing. One may wonder why an apparatus with a 200-ton battery would be needed to duplicate what nature can do in a living room without any special equipment, at least not close by (ignoring the thunderstorm cloud and its electrical engine).

In the same line of thought researchers have used electric welding equipment on silicon wafers to produce time spheres of molten, burning silicon, but again, these are only look-alike BL events bearing no relation to real BL objects.

Another researcher trying to re-create the submarine fire balls was Robert Golka, well-known for his almost life-long involvement with BL experiments. Unfortunately, he died in 2018 at the age of 80.[5]

At the Third International Ball Lightning Symposium at the University of California in Los Angeles, he presented a video tape that showed luminous spheres of molten metal floating on the surface of water. These spheres had been produced by a short-circuit between electrodes submerged in the water. Once again, these were only pseudo-BL. Bob Golka did other experiments to create BL, and these were on a really huge scale.

He single-handedly built a large Tesla transformer in an old US Air Force hangar. It worked well, creating huge bundles of streamers extending from the HV terminal of the transformer, but they were only the normal, thin and filamentary variety of discharge. There is a rumor that once or twice in all the years that he worked with the Tesla transformer he managed to create luminous globes reminiscent of BL, but there are no photos, videos, or other information.

[5] For an obituary see: http://rdmenzies.com/2018/05/04/remembering-robert-golka/ (September 2018).

A Tesla transformer can produce voltages in excess of 1 MV, but it operates at a high frequency, typically 1 MHz. The efficient production of HV is due to the resonant operation of the transformer. A Tesla transformer will produce a large amount of space charge, but the discharges are thin filaments and it will be almost impossible to form a negative space charge of the required shape for the production of localized electromagnetic structures. Nevertheless, it is conceivable that in certain parts of such a huge structure a favorable situation might arise from time to time, but it is obviously not a copious producer of BL objects.

On the internet there are claims that other people[6] have been able to identify the precise procedure Tesla used and imitate it to create BL objects, but I was unable to find any conclusive evidence that that was really the case. It seems that there exist no photos or videos as proof of these experiments, and certainly they have never been repeated by other researchers.

From the accounts of alleged BL production, one gets the impression that almost everything that is luminous and roughly spherical has somehow been linked to BL. The list given above is not even complete, since luminous objects observed in high and ultra-high vacuum have also been compared to BL. Obviously, most of the time there was no proper reality check to see whether what was observed was compatible with observational reports of BL. There may well have been some real BL among those objects, but the almost complete lack of documentation makes it impossible to recreate these situations.

The planned experiments have also failed to produce anything like BL; but none of these have tried to simulate the situations under which BL creation has been observed in Nature.

Before we close this chapter, we shall take a closer look at one particularly interesting type of BL experiment which appears to be quite popular with amateur scientists: DIY BL in a microwave oven. There are several videos on YouTube where brightly luminescent plasma clouds or plasmoids are created in microwave ovens. Of course, these ovens do not normally create plasma discharges in ambient air, this would burn the food instead of heating it, so some tricks are needed to get it started. Most often, something burning is introduced, a burning candle or burning toothpicks. When the microwave power is then switched on, the electrons created in the hot flame gas are accelerated by the electric field of the microwave and they heat or ionize other air molecules. If enough free electrons are introduced as a starter, a discharge is initiated that becomes self-sustaining, and a plasma blob is created which

---

[6] The people in question are and J. and K. Corum.

contains hot, luminescent gas, ions, and free electrons. The plasma blob is not usually spherical, but rather irregularly shaped. Because of the buoyancy of the hot air, it rises like a hot air balloon and hits the top of the oven, where it will burn the paint. Therefore, clever experimenters used inverted glass jars to contain the plasmoid. Still, if the power is left on for too long, the glass may shatter because of the intense and uneven heating by the plasma; therefore, better heat-resistant Pyrex glass is often used. Of course, these plasmoids are not BL objects, but these experiments show us several important things.

First, they demonstrate that, in order to start a discharge in air when the electric field is below the critical 3 MV/m, an injection of free electrons is required. This applies to the microwave experiments where flames are used,[7] but also to the start of normal lightning where the electrons of an air shower due to a cosmic ray particle are probably needed to start the relativistic run-away. For the creation of BL, electrons from a negative space charge are also needed.

Second, the plasmoids have a strong tendency to rise because of their buoyancy, a characteristic which contrasts with observations of BL objects. This in an objection to the plasmoid models of BL which has often been voiced. BL objects appear to consist of very little hot plasma, and the buoyancy must be counteracted by other forces that keep these objects close to the ground.

The third observation is that the plasmoids cannot penetrate the glass of the jar, except by shattering it. The electric field of the microwaves is not strong enough to create a new plasmoid on the top of the glass jar. BL objects penetrating glass panes are stripped of their plasma on the outer side of the glass, so they have to recreate it on the other side. One possibility is that the electric field at the periphery of the BL is parallel to the surface, so that discharges can be started along the surface of the glass. Such discharges are easier to initiate then discharges in open air.

We have seen that these microwave oven experiments do not produce real BL objects, but they are nevertheless quite useful in helping us to understand some of the issues involved in the discussions around this subject.

As a final remark, despite the numerous attempts to make BL, no setup appears to have been used that might have been able to create localized EM structures; so, there is a considerable potential for success if we follow this route. This may be especially true if we try to make a small BL object with a low energy content of, say, 100 J or less, which should be reasonably easy.

---

[7] Sometimes other means like carbon felts are used; see http://amasci.com/tesla/bigball2.html

Summary

- The focus on localized electromagnetic structures allows a definition of the essential requirements for BL experiments.
- First, free electrons are needed since they can oscillate.
- Second, they must form a space charge of suitable shape for an antenna.
- Third, energy input is required to drive the oscillations producing the electromagnetic structure.
- Positive CG lightning near ground provides the negative space charge, while the energy input can be due to the final HV pulse or the EMP of the return stroke.
- Negative CG lightning can produce a negative corona once the return stroke deposits negative charge in the ground or in conductors, but not during the stepped leader stage.
- From negative lightning channels up in the air, production is possible in the corona sheath, but with low probability.
- The pulse equipment used in high voltage labs, like Marx generators, cannot simulate the temporal structure of the electric field due to an approaching leader; the rise time of the pulse is simply too fast.
- The test equipment in HV labs is not designed to produce negative space charge clouds, while avoiding the production of streamers.
- Accidental production of BL-like objects has been reported fairly often, but the replication of the experiments is almost always impossible due to the lack of information.
- So far, planned experiments have failed to produce BL objects.

13

Wrapping It All Up

We have now reached a point where we can ask ourselves whether the BL puzzle is complete: have we fitted all observations and the remaining theory into a convincing picture that leaves no room for alternative explanations? Well, the picture on the puzzle looks good: we have a singular means of BL creation which occurs when negative space charge of a suitable shape, constituted by free electrons, gets an energy burst causing it to emit radiation and create an object which is a localized electromagnetic structure. Indeed, electromagnetic theory predicts objects that could be the origin of ball lightning. It may very well be that the solution to this puzzle was hidden in plain sight for about 150 years. Currently, this is only a hypothesis, but it has a big advantage: it can be tested both by observation and by experiment.

With respect to observations we have now a good idea where to look for BL: positive CG lightning is the best candidate to observe these elusive objects. In particular, thunderstorms that produce a lot of this type of lightning, for example thunderstorms which have ingested smoke from bush fires, are good choices. Singular, very strong positive strokes may also produce a multitude of these objects. Another place to look for BL are very violent volcanic eruptions that generate extensive ash cloud with intense electrification. Storm chasers and volcano enthusiasts have good chances to see and record one of these objects.

When recording thunderstorms and potential BL objects, it is essential that the video or photo should be of high quality. The web abounds with videos that show a wobbly, blurry picture of a luminous but remote object, and it is of course impossible to tell what the origin of the object was. Another important piece of information is the precise time of the observation. Lightning

© Springer Nature Switzerland AG 2019
H. Boerner, *Ball Lightning*, https://doi.org/10.1007/978-3-030-20783-0_13

location systems record the time of strokes with millisecond accuracy, and modern smartphones or cameras with GPS receivers should also be capable of a similar precision. Good timing is important if the observation is to be correlated with the data from lightning location systems.

Experiments should try to simulate the conditions generated by strong positive CG lightning close to the ground, because we have reliable reports on multiple BL production for these situations; the conditions must have been very favorable.

One word of caution: it seems that amateur scientists like to do high voltage experiments with Marx generators or Tesla coils. If you plan to produce BL objects with such an experiment, make sure you know what you are doing and don't get killed in an accident; a dead amateur scientist is a bad scientist.

Odds and Ends

There are some points I would like to mention but which did not fit into one of the other chapters.

Since the 1950s, there has been a hope that somehow nature has found a way to make a stable plasma object which occurs as BL, and that this would provide a way to produce controlled nuclear fusion. If the hypothesis of a localized EM structure is correct, this is of course not the case. The EM structure will expel the plasma from its core by radiation pressure, which is quite the opposite of confinement. It is therefore extremely unlikely that BL research will lead to a new way of achieving nuclear fusion.

Can BL exist in the atmospheres of other planets? For BL we do of course need linear lightning as a starter. As of now, we are only sure that lightning occurs in the atmospheres of Jupiter and Saturn. Both of these planets have water-based clouds that may be vital for the production of lightning. There are some indications that lightning may occur on Uranus and Neptune, but the evidence is not clear. Of course, the atmospheres of these Jovian planets are completely different from the terrestrial one. Lightning is exclusively intracloud, since the Jovian planets do not have solid surfaces like the terrestrial planets. Moreover, oxygen is completely missing because it a reactive gas which is produced on Earth only by the photosynthesis of plants. Therefore, electrical discharges will behave differently on these planets, because there is no electron scavenger. Hence, free electrons will be much more frequent in these atmospheres than on Earth. It is therefore conceivable that similar conditions for BL creation will exist close to the strong cloud discharges observed on these planets.

What is the relation between ball lightning and bead lightning? Bead lightning probably refers to two different phenomena. Some lightning channels will break up into individual segments when decaying, and these segments can look like luminescent beads. The size of these luminescent regions is rather small. There are now several videos on the web that show this phenomenon. I have witnessed another phenomenon that is similar but not identical (see Case 25). Here the lightning channel had a stroke-dot appearance and the bright regions were many meters long, as judged from a distance of several kilometers. The same region of the thunderstorm cloud produced at least four of these strokes over a period of half an hour. A similar behavior is described in a textbook on storm electricity (MacGorman and Rust, 1998): "We and other observers of large storms on the plains have seen that some storms produce many bead lightnings." If you see such lightning, it is therefore a good idea to start taking a video of the region where it appeared, since it is likely that more of the same kind will occur.

Why didn't I do my own experiments on BL since I am so convinced of the correctness of the hypothesis I have given? Well, the ideas concerning the experiments presented in this book are very fresh; in fact most of them have arisen over the last 12 months. I simply didn't have time to try any experiment of my own, and to be honest, I have no experience with HV electronics. The only HV gadgets I have ever built were some electrostatic machines like Wimshurst and friction generators. It would probably be better if somebody with more knowledge of high voltage equipment were to try this out.

Are there observations of BL that do not fit the hypothesis proposed above? Yes, there are observations that are rather unique and are therefore difficult to place in any context. See Case 26 for an example, which is due to Wilfried Heil, with whom I have been collaborating on BL for some time. When we get a reasonable candidate for BL from experiments, we will have to take another look at these reports to see if they can be explained along the same lines as other reports, but so far this is not possible.

Since 1999, when Stenhoff published his book on BL, many things have changed, especially concerning the internet, the general availability of very powerful digital cameras and video recorders, and also the fact that worldwide lightning location systems can now be run by amateurs.[1] Stenhoff expected the BL puzzle to be solved within ten years, but this turned out to be far too optimistic.

---

[1] See https://www.lightningmaps.org

So, when can we expect to have a definite answer to the puzzle of BL? Instead of an estimate of the time needed, I will give you three scenarios, so you can chose the one that appears most likely to you:

First, we have the "business as usual" scenario. There will be no further actions by other scientists to collect observations or to do experiments, and only inconclusive evidence will continue to be collected by amateurs. No progress, or very little, can then be expected over the next ten to twenty years.

Second, we have the "lucky observer" scenario. By pure chance, one of the lightning researchers captures a video that unequivocally shows the creation of a BL object by a linear lightning strike. Other scientists are convinced that this is a real case of BL and start working on the subject. The details of the recording allow estimation of the circumstances involved in the BL creation, so experiments can be planned. Another twist to this scenario is that one of the rare cases with a copious production of BL objects is witnessed and recorded by several people, and that the sheer amount of reliable evidence rekindles scientific interest in this phenomenon.

Third, the "clever physicist" scenario. A young physicist reads this book and gets a bright idea about how to perform an experiment in some easy way. She builds some apparatus and, after tweaking a few parameters, succeeds in producing small BL objects in a repeatable manner. The objects can be studied in detail and all the open questions can be answered.

I hope that one of the more optimistic scenarios will come true!

Appendix

Case 1: Neuruppin 1994

This is the account of a BL event that took place in the town of Neuruppin, Brandenburg (Germany), on 15 January 1994. The report is based on an earlier publication (Bäcker et al., 2007), but has been updated because additional information became available.

The event is unusually well documented through fortuitous circumstances: the staff of the local meteorological station witnessed parts of the event and the director of the station started to collect eye-witness reports immediately. Supporting information about the lightning activity of the thunderstorm is available from Siemens' lightning detection network (BLIDS). In total, about 11 different BL objects were reported, a very high number for a single event. The amount of information available allowed a clear correlation between a lightning stroke and the multiple BL observations.

Geographical and Meteorological Situation

Neuruppin is a town of about 30,000 inhabitants situated in flat countryside about 60 km northwest of Berlin. The town borders the western shore of a long north-south running lake (Fig. A.1). The meteorological station was at its southern outskirts, on the border of the lake.

On 15 January 1994, maritime air of polar origin was moving at a speed of about 90 km/h from west–northwest over northern Germany. In a trough

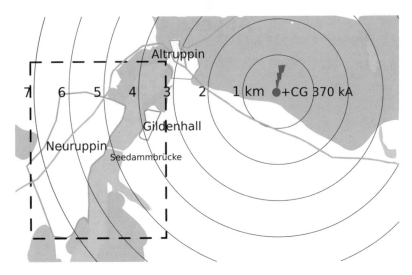

Fig. A.1 Map of Neuruppin with the position of the first positive stroke. Distance circles in kilometers. The area marked with dashed lines is represented in the second map. Drawing by the author

extending from Greenland over Scandinavia to the North Sea, a cut-off process started over Ireland, leading to the formation of the low-pressure zone "GRETE". "GRETE" had its center over Brittany at 0:0 UTC on 16 January 1994 and moved relatively mild maritime air of sub-polar origin towards central Europe. At the same time high-reaching polar air moved southward over the North Sea. A cold front of the low-pressure zone "ELSE" over Finland produced several thunderstorms at the frontier with the milder air mass over central Europe. During 15 January the thunderstorms moved from the North Sea to the greater Berlin area. The following meteorological stations recorded thunderstorms: Helgoland from 6 to 9 UTC, Hamburg and Bremen from 14 to 15 UTC, Bremen, Lüchow, and Neuruppin from 15 to 18 UTC.

The principal requirements for the development of winter thunderstorms (lifting and unstable layering) were certainly present. At 0 and 7 MET at the 500 hPa-level (about 5000 m), a temperature of −34 °C was measured.

At the 850 hPa-level (around 1500 m) the temperature was about −5 °C (analysis of 15 and 16 January, 0 UTC). Therefore, the temperature gradient was about 28–30 K, which is sufficient for the development of showers and brief thunderstorms in wintertime.

The Neuruppin Events

Until shortly after 17:00 local time (16:00 UTC), there was no indication of a thunderstorm in Neuruppin. At that time, it was already completely dark.

The event that then took place is best described in the words of the meteorologist on duty, Th. Hinz: "An exceptionally bright flash of light was seen originating from the north, followed by a very loud thunder clap about 10 seconds later." Later, the staff of the station observed three more discharges in the form of linear lightning. Because the office had no windows facing north, the source of the light could not be seen directly, but all objects in view were brightly illuminated. The weather diary of the station notes: "Thunderstorm (light) N-E 16:06–16:28" (time in UTC).

Soon after the first discharge, telephone calls were coming in at the station, with people inquiring about the nature of the light and the noise, both of which were generally perceived as extremely strong. The director, Donald Bäcker, collected data from people phoning in and he also made an appeal in the local newspaper for more reports on any unusual observation connected with this event. He did not mention BL in this appeal, though. To his surprise, the reports did not only contain accounts of the primary flash, but often mentioned luminous spheres above the rooftops or inside houses. He was at first inclined to dismiss these reports as not being credible, but as more and more reports of a similar nature continued to come in from different people, he became convinced that they were real. In total, 34 reports were collected, which included the identity of the observers and a brief transcript of their observations. After finishing the collection of reports, D. Bäcker published a summary in the local newspaper on 31 March, and in the supplement of the internal report of the German weather service. At that time, he thought erroneously that the primary lightning had also been some kind of ball lightning.

About a year later K. and S. Näther visited all the witnesses to interview them. With few exceptions, people were ready to give more information about their observations. In these reports, five classes of phenomena can be distinguished:

- the primary lightning flash (the light and the loud thunder),
- corona discharges,
- two large BL objects over the rooftops and over the lake,
- several small BL objects inside houses and five around houses,
- a residual category which includes widespread damage to telephone systems, effects on the warning lights of a railway crossing, and similar.

The spatial distribution of the observers' locations is centered at Neuruppin, and covers several square kilometers (see Fig. A.2). There are no reports from villages further east, but some observations were made at the eastern shore of the lake. For a detailed description, see the most important witness reports below.

The lightning detection network BLIDS recorded only five lightning discharges on 15 January 1994 in a square of side 50 by 50 km, centered at Neuruppin (see the table below). The first was at 16:08 UTC, a positive lightning stroke with an enormous peak current of 370 kA. It was followed by four less intense discharges between 16:09 and 16:22, where the two last were nearly coincident. Three of these were also positive CG discharges and there was one negative CG. Thus, the numbers of discharges noted by the meteorologists and by BLIDS agree. BLIDS located the first discharge about 5 km east of the center of the lake bridge. The other discharges were recorded even further east. BLIDS data gives a distance of 7 km between the lightning hit point and the meteorological station, whereas the estimate from the time interval between thunder and lightning as noted above (about 10 seconds) is only about 3 km. The location accuracy of BLIDS is given as 1 km. Data from two cases where positive lightning destroyed trees (Heidler et al., 2004) give a difference between the BLIDS data and ground truth between 1 and 2 km. There is a slight discrepancy here but there may have been an intra-cloud discharge first, which was then closer to the observers than the final CG stroke. In one report (16) it is stated that the thunder started with a rumble before the final, load crash, which may support such an interpretation. Considering the difficulty in judging time differences, one can consider that the data are compatible.

Obviously, all observations of BL objects appeared to coincide with the first, positive lightning stroke of this winter thunderstorm. No linear lightning channel was observed; all observers with a direct line of sight to the measured hit point describe it as a point-like source. The lightning channel must have been much shorter than usual.

Initially, before the BLIDS data were considered, this was a source of confusion between the large BL objects and the primary flash. Another possibility is that a considerable part of the lightning was an intra-cloud discharge, which was screened from the view of most of the observers, so that only the short connection to ground was visible. Some observations may support the existence of an intra-cloud discharge, like number 26.

Positive lightning of considerable strength has been reported repeatedly for winter thunderstorms (Rakov, 2003). Wind shear, displacing the upper, positive end of a charge dipole from the lower, negative one, has been put forward as a likely mechanism for this type of lightning. In the case of Neuruppin, the actual cause is unclear. The lightning detection system recorded a cluster of negative lightning flashes close to the town of Kyritz, about 30 km west of Neuruppin, but these occurred about 18 min later, so it is unlikely that this was the lower, negative charge. From the data it is more likely that an inverted dipole was involved: the lower, positive end of the charge dipole caused the positive flashes at Neuruppin and further to the east, where the last one was coincident with the discharge of the upper negative charge. In addition to these two lightning clusters, the thunderstorm produced only two other flashes in a square region of 150 by 150 km centered at Neuruppin.

Several observers noted corona, one on a very large scale: observer 5 on the eastern shore of the lake, with a free field of view towards the hit point, observed huge bundles of blue flame extending to the sky. Observers on the western shore also reported corona discharges, but of a more common scale, e.g., blue light appearing on a metal sieve being used for sifting sand (observer 7) or blue sparks emanating from the microphone of a telephone handset.

It appears that observer 5 saw negative leaders, indicating an electric field in excess of 3 MV/m. This is roughly the value obtained in high voltage engineering for the breakdown field of negatively charged conductors. The corona discharges were present for a period of several seconds (observers 7 and 13).

Coincident with the primary discharge, several objects were observed which fit the definition of BL: luminous, floating, spherical or elliptical objects with a lifetime of several seconds. The largest object was described as a bright, yellow ball of about the size of the full moon, hovering motionless over the rooftops. About four observations can be ascribed to this object. A second large, red spherical object was seen passing over the lake bridge from south to north and vanishing in or slightly above the lake; for this object there are at least three independent observations.

At least 9 smaller BL objects were seen outside and inside houses. In four cases, more than one observer saw the same object. Several of these objects occurred only briefly, for about a second. This is the case for two of the objects seen inside houses, where the BL objects were observed from appearance to extinction. Other BL objects existed for longer, so they could be observed with ease; especially one which was seen between a Yagi antenna and the metal railing of a balcony (number 23). Some of the objects showed the type of independent motion typically associated with BL. Generally, the shape was described as spherical, but in one case, where the object could be observed over a longer period, the shape was described as round but flat, like a satellite dish. The witness saw this object through a window less than a meter away and saw its shape shifting. Another object was described as "oval like an egg". Obviously, there is a considerable spread in the characteristics of these BL objects, but the evidence nevertheless points to a common origin, being linked to the exceptionally strong positive lightning stroke.

Skeptics of BL observations often claim that they are afterimages on the retina due to exposure to the intense flash of light from the linear lightning.

However, none of the witnesses observing the smaller BL objects had been able to see the primary flash directly, so this explanation can be excluded. Since the strike point of the primary lightning was several kilometers away from most of the observers, optical illusions like phosphenes due to the exposure of the brain to strong and varying magnetic fields can be excluded as well. Furthermore, theories claiming that BL objects are due to illusions created by external factors like light flashes or magnetic fields will find it impossible to explain the consistent observation of the same BL object by several people, and also the fact that one of the objects slightly singed a curtain. One has to conclude that the observations are based on real, physical phenomena.

The considerable distance between the observations and the primary flash presents an insurmountable obstacle for all theories requiring the direct interaction of the lightning channel with the material forming the BL. This relates to theories based on combustion of materials like carbon or silicon.

At a distance of several kilometers, the primary flash can have acted only via the electric field or the strong electromagnetic pulse (EMP) which is often associated with such a strong lightning stroke. Therefore, the strength and maybe the duration of the electric field must have played a central role in the copious production of BL.

Several of the smaller BL objects were reported to have moved through curtains or—in one case—through windows. In another case, the curtain is reported to have had brown spots at the position where the BL penetrated it, the other curtains and windows were undamaged. All BL models based primarily on physical matter will find it impossible to explain such behavior. The only models which appear to be compatible with all the facts are those exclusively based on electromagnetic radiation.

In summary, the Neuruppin case stands out from other BL observations because it is very well documented. A first round of information was collected immediately by a meteorologist and more detailed information less than a year later. The sequence of events makes it very unlikely that several people conspired to create a hoax by producing fake observation reports. The observations can be well correlated with data from the lightning detection network BLIDS, enabling a satisfactory reconstruction of the events by linking the first lightning stroke of the thunderstorm to the observed BL objects.

All evidence points to the fact that the many BL objects were created several kilometers away from one single, very strong positive flash, in a region where the electrical field was at very substantial levels. Based on this evidence, BL models calling for a direct interaction of the lightning channel and the material forming the BL object can be excluded. Moreover, BL models based on optical or other illusions associated with the light or the magnetic field of the primary flash can be excluded with certainty.

Since the primary flash was not the direct source of energy for the formation of BL objects, one is forced to conclude that other and weaker sources, like the energy stored in the electrostatic field, must have been involved.

Date	Time (UTC)	Longitude	Latitude	Type	Strength (A)	Distance (m)
15.01.94	16:08:36.698	12.890	52.940	CG	370,000	5900
15.01.94	16:09:32.481	12.905	52.944	CG	34,000	6900
15.01.94	16:10:51.260	12.929	52.945	CG	97,000	8500
15.01.94	16:22:11.961	13.119	52.950	CG	24,000	21,000
15.01.94	16:22:11.979	13.014	52.885	CG	-89,000	16,000

In Fig. A.2 the reports on corona are marked blue, whereas reports on BL are marked red. Some of the BL reports have not been marked.

Witness Reports

From the 32 reports, only those concerning BL observations or observations relevant to the lightning and its effects are listed here.

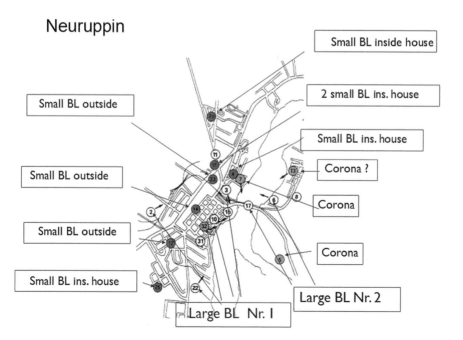

Fig. A.2 Positions of the witnesses in Neuruppin. Drawing by the author, based on a drawing by K. Näther

Witness 3

"I drove my car into the garage and went home. When I was at a railway crossing, I was suddenly bathed in bright light. The whole area around was brightly lit. It was completely soundless. The intensity of the flash could be compared with the flash from a radar speed check, but here it completely illuminated the surroundings. It vanished as suddenly as it appeared, and I went on my way, but it left quite an impression. Two or three seconds later there was a very loud bang. To me, the whole situation appeared very strange. I met nobody, I was completely alone."

At the same time, the wife of this witness was sitting at the table in the living room doing same handicrafts: "The door to the balcony was open, but the curtains were closed as usual. Suddenly a bright ball came flying into the room through the window and vanished immediately. I was frightened. It happened so fast that I could not see the color or anything else remarkable. Soon after that, there came an extremely load bang from outside, which really took me aback. Until then it had been a quiet evening. Outside there was nobody who could have thrown something through the window. It never occurred to me that this had anything to do with a thunderstorm."

Witness 4

Mother, son, and daughter were sitting in the kitchen eating pizza. The window behind the closed curtain was canted in order to let fresh air into the slightly smoky room. The daughter was sitting with her back to the window and looked towards the stove. The son was sitting opposite to her. The mother was in the middle of the room.

Daughter: "We were all sitting at the table. Suddenly I saw a tail vanishing in the direction of the stove. I was immensely frightened. Through the open window came a slight draft of air, and the object seemed to follow it. Shortly after that there was a bang, as if a plane had broken the sonic barrier just above the house. I was trembling all over, I could not understand what was going on. Our small boy was watching TV in the living room. I immediately checked to see what had happened, but everything was OK."

Son: "I was just about to start eating when, from the height of the window, a very bright ball or tail approached me. It was the size of a hand ball and white, and certainly not yellow. I could not see where it went, it was moving too fast. Despite its speed, I could recognize its spherical shape. I was shocked. Immediately afterwards there was an enormous bang. The cups in the cupboard were shaken! I had never witnessed anything like that in a room. We called the weather station to find out what had happened."

Mother: "I was sitting at the table, when I was surprised by something bright. Immediately, my daughter started to scream. It was not raining outside, so I did not think of a thunderstorm when it cracked outside. Maybe a plane was flying over us, we had heard sonic booms before. The electrical appliances in the kitchen were undamaged. We only found two brown patches on the cotton curtain of the kitchen. Unfortunately, we have already thrown it away."

By chance a neighbor came in during the interview and told us what she had experienced that evening: "I was sitting on the couch, when suddenly I had the impression that all was falling to pieces behind me. My body was thrown back and forth. The whole courtyard was brightly lit. A short time earlier I had been outside, and the sky had an unusual appearance" (Fig. A.3).

Witness 5

"At that time, this house was only a weekend cottage. And the other houses were not there. We had an clear wide view in an easterly direction. I am not afraid of thunderstorms, I like to watch them. On this day, I was working in the kitchen. Dark clouds came up which I kept watching constantly. Suddenly, there was an unusual flash. It could be compared in appearance to a poorly adjusted electric welder, which produces similar blue sparks. But what happened out there was on a much bigger scale! Huge blue bundles were stretching towards the sky. When I heard a loud bang, I thought that something had

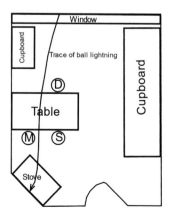

Fig. A.3 Situation of witness 4. Drawing by the author, based on drawing by K. Näther

happened with the HV masts. The light and the bang were unusual and not typical for a thunderstorm. Only much later I heard a grumbling noise. A thunderstorm in winter?"

Witness 6

"Though it had rained a bit, there was no indication of a thunderstorm. Suddenly it turned incredibly bright, without any sound. I saw a large, red fireball flying over the 'Seedammbrücke' in the direction of Altruppin. It descended somewhat, so I could no longer see it from the window. Maybe it dived into the water or it dissolved. I had the impression that somebody had shot a big balloon that was moving without sound over the water."

Witness 7

"On this Saturday afternoon my wife and I were sifting sand in the yard. It was a nice day, quiet and mild. At about 5 o' clock I got the impression that it started to drizzle. The sand sieve was lying on a metal wheel barrow, its wooden frame extending over the edge of the wheel barrow. Suddenly, the sieve was lit up in a blue light. In my 40 years I have seen many things, but that gave me quite a shock. The light vanished, and I wanted to continue working. Suddenly there was a loud bang. My wife went inside, but before I could make up my mind to follow her, the sieve was lit up again in a blue light. Now I was worried, because the whole thing clearly had something to do with electricity and that should not be taken lightly. The lamp in the yard was not destroyed. When my wife told her mother what had happened, she said that she had seen something weird flying over the lake and diving into it. We contacted the weather station only after we read the call for witnesses in the newspaper."

Witness 8

"I was driving in my car with my daughter from Alt-Ruppin towards Gildenhall. Quite close to the horizon we saw a white ball, which dissolved itself while throwing out sparks. At first, I thought it was a speed control by the police, but 3 seconds after the disappearance of the sphere, we heard an ear-splitting bang, which frightened us very much. At the same time the warning lights of a railroad crossing came to life, but there was no train on the track."

Witness 10

I was driving down the Klosterstraße with my wife. I turned left for the Poststraße. It was already dark. We had passed down about half of the hundred-meter-long road when we noticed, just above the rooftops, a blinding white sphere. It did not move and was about the size of a soccer ball. The object had sharp contours. I did not have to lean forward in the car to be able to see it through the front shield. I had the impression that we were not far away from the phenomenon. I estimate its height above ground to be 50–100 m. Then the object vanished, as if switched off […] A few seconds later—we were already driving though the Poststraße—there was a loud thunder clap. It had not been raining."

Witness 12

"We—my friend and I—were biking home from the indoor swimming pool. It was already dark and the street lights were on. Near the Junkerstraße there is a combined pedestrian-bike path. To the right are new buildings, to the left a shoe shop. Suddenly there was a loud bang which frightened us. Then something round, whitish or yellowish, approached us from above. It touched the ground and was gone without any sound. At that moment one streetlight went out. Now we were even more frightened, and we hastened home."

Witness 13

"I was climbing the basement stairs when I noticed a bluish glow above our garage. It formed a semi-circle, wider than the six-meter-long garage. In total it may have been 15 m wide, and it was about 2 m higher than the roof of the garage. Around the semi-circle there was a crown of light rays. The whole spectacle last several seconds, so I could gaze at it unhurriedly. Shortly afterwards there was a terrible bang. Even my dog, who is accustomed to gun shots, was made to jump slightly. I thought that maybe a thunderstorm was coming up, but it remained quiet afterwards. The bang I considered untypical for a thunderstorm."

Witness 14

"While talking to an acquaintance on the phone, which is in the corridor, I looked through the closed window into the small inner court. I was half a meter away from the window. Suddenly, left of the chimney, there appeared a silvery bright light. I couldn't understand where the light was coming from, and then the object approached me. I was afraid of being hit, yet the closed window comforted me; but what if it had been open? The object came down from the roof in an oblique path. I noticed a type of trail behind it, but that may have been an optical illusion produced by the brightness and the speed of the phenomenon. It approached the window, so I could see it very well. It didn't have the shape of a meatball at all, but was flat like a satellite dish. The object was less than a meter away, 2/3 of it under the window sill, the rest clearly visible. It didn't have a rigid shape. Sometimes it changed its shape slightly, as if unsure in which direction it should go. The whole thing lasted for a while, at least as long so I could tell my acquaintance on the phone that there was a funny type of lightning. Then, to my relief, the dish vanished vertically upwards. There was a very loud bang, which made me think of a thunderstorm. I did not feel heat or anything similar, but there was a crackling noise on the phone."

Witness 16

"On that day, we were all at home: my two children, my husband, and my mother-in-law. I was sitting in my favorite corner of the living room, with my daughter playing around near the armchair. Suddenly, I looked up because somewhere there was a bright light between me and the child. At first, I thought that the child had been playing with the camera, which was lying on the armchair. I was about to scold her, but she looked at me with a frightened expression. From the hallway, my husband was also looking at the light with a stunned expression. It lasted only 1–2 seconds, enough to take it in and notice that my daughter had not used the camera flash. It was directly over the floor, oval-shaped like a large egg and about 1 m long. It was blue in the middle and very bright around the outside. It didn't move, and it vanished just as suddenly as it had appeared […] I thought there would be a scorch mark on the carpet but could find nothing. Nothing in the house was damaged […] We did not have time to react in any way, I just looked up. I was breathing heavily. Outside there was a rumbling and then a very loud bang. It sounded as though a big lorry had toppled over, skidded over the crossing, and then crashed into something. We all looked out at the crossing and saw the people from opposite also looking out. My husband said that there had been a strong thunderstorm outside, but I do not think so, because it was completely quiet."

At the same time the son saw a similar object in the neighboring room, close to his sister's bed, but it was a yellowish color (Fig. A.4).

Witness 17

"With a friend we were driving by car on the "Seedamm" bridge towards the city center. Something very bright was coming towards us. Everything around was so brightly lit that we couldn't see anything through the windows. Probably the object went directly over the roof of the car. There was no traffic in the opposite direction. We could only see that to the right it either disappeared into the lake or it simply dissolved. Only then did we stop the car. We got out to recover from the shock. I accidentally touched the roof of the car. We had been on the way for 2 minutes and the air was cold. Nevertheless, the roof was definitely warm. The car heating could have had nothing to do with it. Suddenly there was a loud bang, which even my mother heard at home. I got the idea that this might have been a thunderstorm and an encounter with an unusual kind of lightning, and I called the local weather station. I was surprised that our car radio, which had been switched on, was undamaged. For us it was a scary encounter."

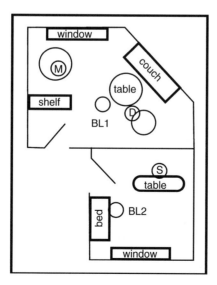

Fig. A.4 Situation of witness 16. Drawing by the author, based on drawing by K. Näther

Witness 20

"At that time, we–my son, a friend, and me–were in the workshop. A wood milling tool was standing there, about 80–100 cm away from the outer wall. Suddenly, something as bright as a magnesium torch was lit up in this space, the workshop was bathed in a bluish light, strong like sunlight. The luminous object was about 1 m above the floor, and it had a diameter of 50–60 cm. My friend was facing in the opposite direction, so he could only see the light, but my son ran towards me and hugged me. He had seen the ball of light. The duration of the whole thing was very short, hardly more than a second. Soon afterwards there was a loud, scary bang which pressed us against the wall. Until then it had been quiet. My son is sure that it started to rain shortly afterwards. We checked all machines and electrical appliances in the room. Nothing was destroyed, everything was working well even though all the power plugs had been plugged into wall sockets. All the windows in the room had been closed."

Witness 23

"I was sitting on the couch doing some handcraft. I was able to look out of the window in a north-easterly direction. Suddenly there was a luminous ball on the balcony, directly in front of the window. At first, I thought it was a lost balloon, but it was sitting there, much too quietly, between the metal rail and the antenna (Comment added: a VHF TV Yagi antenna). With a diameter of 40 cm, it just sat for a very long while on the balcony. I don't know if it was more than a minute, I didn't look at my watch. I could see the color well, but it is difficult to describe. It was not a proper red, nor yellow, maybe orange comes closest. It was not blinding, so I was not afraid. I could hear thunder outside."

Witness 25

"I was sitting in the corner of the room reading a newspaper. Suddenly, something hissed past me by. It came through the closed window. The net curtain was closed, but the main curtain was open. The object had the size of a small glass lampshade and it looked like a yellowish light bulb. From the window it moved to the opposite side of the room along the cupboards to my left, then it made a U-turn and came back straight towards me, but luckily it didn't touch me. It passed me at a close distance and went back out through the window. In the first moment of shock I thought that somebody had thrown a stone through the window, but this would have broken the glass pane. It went by very swiftly. It hissed and the room was brighter than if the sun had been shining in. Outside there was a strange thunderstorm. Instead of normal lightning, I later saw a broad horizontal luminous band above the opposite roof. It appeared to come from the left, vanishing to the right. It was strange. I did not hear any thunder. Neither the curtain nor the TV was damaged" (Fig. A.5).

Witness 26

"At that time, I had had the telephone only for a short while. When I make a call, I usually look out of the window toward the north. That day a former worker colleague had called me. She was surprised because there was a crackling noise on the line. I told her that we had a thunderstorm, and she informed me that one should not make calls during thunderstorms. I did not know that, because I was new to telephones.

Suddenly the whole sky turned a vivid luminescent blue, even though it had already got dark. The phenomenon appeared to pulse, sometimes it was stronger and sometimes paler. Then I noticed that blue sparks were flying from the microphone of my handset. Frightened, I asked my colleague if she was seeing the same thing, but she said she wasn't. Then there was an almighty bang, which made me jump and it all vanished."

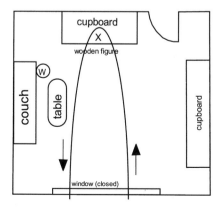

Fig. A.5 Position of witness 25. Drawing by the author, based on a drawing by K. Näther

Case 2: Melbourne 2002 by Ern Mainka

Ern's Original Account

Weather

Weather on the day, 1/2/02—Mid 30s, very humid with late thunderstorms. High temperatures and humidity on previous days. From a Bureau of Meteorology media release: Melbourne recorded 6 thunderstorm days for February 2002—the highest number on record. It was also the coolest summer since 1995/96. The temperature did not reach 100 °F this summer. Rainfall was above normal for summer. This was the fifth year of a drought period with yearly rainfall totals well below normal.

Camera Type and Film Used

35 mm Nikon with 35–105 mm Nikkor short zoom set at about 50 mm. Fuji Velvia 50 ASA transparency.

Scanning

On Linotype—Hell drum scanner. Scanned by 'Splitting Image', Unit 3 Monash Corporate Centre, 10 Duerdin Street, Clayton, Vic 3168. Ph 9542 7400.

All images other than BL frame scanned on an Imacon Flextight Photo scanner. Prints of these images are available. All images available as 300 DPI TIFF files.

On 1 February 2002, during one of the most severe thunderstorms seen in Melbourne, Australia, for many years, I observed and photographed a rare type of lightning, ball lightning.

As well as being a significant event in itself due to its apparently large size, the rarity of existing photographs and the number of eyewitnesses located (seven so far) makes it more so. The number of existing photographs of ball lightning are few.

The ball lightning was captured toward the end of time exposure of a few minutes' duration. This was the first of five consecutive frames taken of this particular storm cell. I also observed it visually (not through the camera lens) from the time it reached full intensity to when it descended out of view. The moment before it appeared, I was watching more intense strikes slightly to the right (see photo below) and I cannot specifically recall noticing the strike that it appears to be connected with. It appears that this strike occurred before the ball lightning appeared and may perhaps be unrelated. Scientific opinion also suggests this to be the case. However, at other times it has been observed to be 'striking off' cloud-to-ground strikes. I have seen BL perhaps three times before in Melbourne. Twice out of those three events I saw it striking off normal CG strikes. It seems it can occur in either circumstances.

By seeking other witnesses to this event through newspaper and television media I was able to piece together its journey of 12 km from Wantirna South to Blackburn. Before it descended it appeared to be almost stationary or hovering for a moment. It then suddenly dived at around 720 kph or more to its first impact point in Vermont. The outer corona diameter appears to be perhaps 125–150 feet (45 m). It divided after the first impact into smaller spheres "with interconnecting color to give almost the impression of part of a necklace or string of beads"—Joe Stanley (Fig. A.6).

It became a quest for me to find out and as a result of news coverage of this the event, I've had seven reliable witnesses contact me. One in East Camberwell (10 km west of the BL), two in North Dandenong (12 km S/E), one at Nunnawading, one in Vermont, and two at its final destination at Blackburn. I was surprised at the difficulty in finding more witnesses, given its size and feel sure there are other people who saw it. More may still contact me. Their descriptions would be invaluable.

The sequence of events that were witnessed at around 9 pm on 1 February 2002 follows:

1. It first appeared over the Wantirna South area. It seemed to appear independently of any lightning stroke. It took 2 seconds or more to fully brighten. It appeared to hover or, as the photo suggests, move very slowly for a further 2–3 seconds. It then suddenly descended very rapidly over 4–5 seconds until it was out of view. It then impacted objects or terrain at least twice, 'bouncing' 12 km to Blackburn. It perhaps disintegrated a little each time as it struck obstacles, while trying to follow the terrain horizontally westwards—even it seems rising over higher land.
2. The first impact point at Vermont triggered or was simultaneous with the large jaggered strike that is captured in the next frame on the roll. (I recall the motordrive advancing the film to the next frame quite soon after the event and then I immediately opened the shutter again via a rubber air cable release that I use in thunderstorm shoots). The ball or at least part of it appears to have clipped trees, killing one medium sized tree and slightly

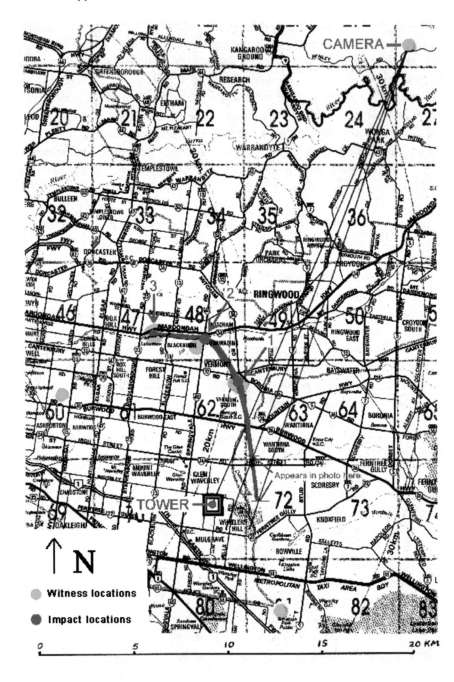

Fig. A.6 Map of BL track according to observations. Drawing by E. Mainka

damaging another in Vermont. A loud explosion was heard followed by a large (30 m) orange glowing light which persisted among a grove of trees for about 5 seconds. It dimmed slowly over a few more seconds before disappearing completely. "The light had a strangeness about it, but looked something like as if a distress flare had been lit. I nearly hit the roof when it struck"—Paul Maher.

3. During the second stage of its journey it was observed from East Camberwell as consisting of four spheres in line. The second was possibly a bit smaller, with two smaller ones again following closely in train, with perhaps a short tail at the end "to give almost the impression of part of a necklace or string of beads. They were connected by colour and lightning" said J.S. "They appeared gold/yellow at their centres with dimmer orange/red flashing corona around them".

4. It was then briefly glimpsed passing over the area of Nunnawading Station—with another loud bang occurring. Very roughly estimated diameter of the orange glow is about 40 ft (12 m) and height of 30–40 feet. The witness a few blocks away thought that a petrol station had exploded.

5. It was finally seen 'rolling' off a railway embankment at Blackburn. Its 'tail' still extended up the slope (~20 ft) when it seems to have contacted a metal BBQ plate and disappeared instantly with a very loud bang, leaving no trace. The yellow ball at the head was estimated to be one metre wide and filling the backyard with an orange/red glow.

I estimate the total duration of the entire event lasted around 35–45 seconds.

Individual Eyewitness Accounts

Summary

Joe Stanleys' observation of it hurtling through the air in an upward curving path (positive parabola) indicates it 'bounced' or was 'propelled' some hundreds of feet in height, travelling 5–6 km to Nunnawading/Blackburn after the first apparent impact point at Vermont. The "almost simultaneous [...] jaggered path flash" seen by J.S. was most likely the cloud-to-ground strike occurring at the first impact point that was captured in the next frame (see below).

At its second impact point - around Nunawading Station - another loud explosion resulted and it was observed fleetingly to have an outer corona of "about 40 feet" (B. Steventon). It then followed the train line over or through Blackburn Station and toward Laburnum Station, but not quite all the way.

Finally it failed to take a bend on the line and ran off an embankment some 12 km from its starting point and simply disappeared in a very loud explosion. Somewhat smaller by that stage, its central yellow core was seen to be about one metre in diameter and filled the backyard with its red/orange corona light.

I conclude that this was the sequence of one continuous event. Most witnesses recall the exact date and approximate time, or for others "around 9 pm on the Friday night when we had that huge storm at the beginning of February", which nearly everyone in Melbourne remembered. Unfortunately, I have not been able to find out the precise time (to the minute) that it occurred. All the witnesses concerned contacted me after either seeing the TV news report or the following local newspaper article in which I invited anyone who saw it to contact me. I believe many more must have actually witnessed it (such as at the busy intersection of Springvale Road railway crossing) and more may still be found. Including myself, there are presently eight known witnesses.

1. North Dandenong—Fay and Peter Barnes (About 12 km from BL)
"Looking NW to NNW an orange ball appeared, slow at first and then descended down, slightly to the left if anything" Peter said. It was observed low in the sky through a gap between trees. A clear view of the true horizon was obscured by more trees. It looked unusual enough for Fay to quickly draw it to the attention of her husband; they were both watching the intense lightning storm at the time. Her thoughts were that it was possibly a plane on fire and expected to hear reports, but there weren't any.

These witnesses were to the south and were very useful in enabling me to complete a triangle of observers surrounding the event.

2. Vermont—Paul Mahar
"A very loud bang preceded a large (30 m) orange/red glow which persisted for 5 seconds and then faded over a few more seconds. It was in a patch of trees in an adjoining paddock. One tree has died. It looked something like as if someone had lit a flare", Paul said.

Two trees (50 m apart) seem to have been hit. The first one showing a circular area of dead leaves as if a sphere had passed through it with the leaves at the top still green. They were not burnt, just dead. The second tree—a 20–25-

ft eucalyptus—was near the center of the glowing light that was observed and shows strike damage (typical of a normal strike) and is completely dead. There is no fire-like charring, but just fine shredding.

It looks very likely that the massive strike captured in the next frame was somehow triggered or connected with the BL touching down here. I seem to remember the motor drive advancing automatically relatively soon after the BL observation and thinking 'phew—got it'. The strike is also positioned in the correct vicinity of Vermont. I released the shutter for the next frame instantaneously where this strike is captured and seems to fit with J.S.'s description of a "jaggered flash" that was almost simultaneous with the appearance of the spheres.

3. East Camberwell—Joe Stanley, Gavan St. (About 5–10 km from BL)
"On the night of February 1st, 2002 between the hours of 8.45 pm and 9.15 pm I witnessed an electrical storm.

Through the glass doors and windows on the northern side of my home I noticed severe and extensive lightning. In particular I noticed a severe and jaggered pathed flash. Accompanying the flash almost simultaneously I saw lightning of a yellow/orange/gold color, composed of what appeared as spheres with some interconnecting color to give almost the impression of part of a necklace or string of beads. It was not connected by lightning, either to the cloud or the ground.

I recall seeing four spheres in the streak. Of the four balls, the first two appeared to be both larger, brighter, and more colored. The path of the 'lightning' was not jaggered as in a natural lightning flash, but followed a slight parabola. It was not accompanied by a thunder clap which I had expected to hear.

The path of the parabola appeared to come from the north east and was headed away from me in a north westerly direction. My first observation of the spheres occurred in a direction approximately 30° east of north.

From my observations I thought that the lightning might have terminated near the home of my daughter near the intersection of Middleborough and Whitehorse Roads. I rang her to see if there had been a lightning strike in her location and she informed me that there had not been. Considering my observations, I found that surprising". Joe Stanley

I recreated an image of Joe Stanleys' observation after its first impact at Vermont and heading towards Nunnawading.[1] Sparks or lightning were also

---

[1] The figure made by Ern Mainka is not included here.

observed emanating from it and it was changing shape which is difficult to illustrate. Joe also observed an uneven outer yellow tinge on at least the first ball which appeared to change shape.

4. Nunnawading—B. Steventon, (about 100 meters away, S/E)

Mrs. Steventon saw an orange light briefly in the sky through a window looking toward Nunnawading Railway Station, a hundred meters or so away. It was about 30–40 feet above the ground and about 40 feet wide. A simultaneous loud explosion occurred. She said at first she thought a petrol station had gone up. She could not discern if the glowing orange light was moving as she was watching TV at the time and caught it almost entirely out the corner of her eye and through a window. She ran outside immediately, but it had gone. She commented that the light looked very strange and difficult to describe. However, seeing the photo I'd taken on the 6 pm news and in the local paper provided her with a clear explanation of an otherwise unaccountable event.

5. Blackburn—E & S Hall

A ball of light was observed 'rolling' down a railway embankment and into their backyard. The railway line is mostly obscured by trees from the lounge room where it was observed from. It's core of bright yellow light was about 1-m diameter and lit up the surrounding yard and trees with an orange/red light (corona). The view of it was partly obscured by vegetation but it appeared 'long' as in having a tail trailing back up the embankment. It made a very loud 'whooshing' noise as it came down, stopping at a low-level metal BBQ plate, where it disappeared with a loud bang. There appeared to be no sign of fire even though it was covered in dry pine needles. The event was short, lasting only about 2 seconds. Her teenage daughter was also in the lounge room but was seated and only saw the bright 'glowing' lighting up the backyard just before the very loud explosion.

It appears this was final 'ground zero' and tallies with Joe Stanley's estimate of where he thinks it would have come down from his viewpoint in East Camberwell/Burwood.

The resulting 1 m wide and very bright yellow 'core ball' of the BL was seen to be 'rolling' off the embankment (hidden from view by trees to the right in the picture) and down to the metal BBQ plate. No traces could be found. The dense foliage on the embankment suggests it rather floated over the tangled dense shrubs. However, a forward rotation as in 'rolling' was seen to be exhibited by the ball. There was strangely no apparent damage to the foliage (but it had been four months since the event) and probably skimmed over the bushes.

Case 3: Lamington, Queensland 1922

From (O'Reilly, 1962): "November following the eclipse brought another phenomenon which did not have such happy associations: it was an electric storm. It loomed in the west about 2 o'clock one afternoon. I was on my way to Kerry with the cream pack-horses and Tom, who was off to Cainbable to look over dry cattle, rode with me as far as the turn-off. The storm came up black and nasty-looking, but no worse in appearance than dozens of others into which we had ridden; there was, however, something uncanny about the thunder; instead of the usual desultory boom of a coming storm there was a continuous sound like an endless procession of great steel balls rolling down a long stone corridor. At the cliff top I had a close view of the coming horror and that was enough to send me racing back to Luke's empty humpy for shelter; the clouds were higher than the usual storm and tinged with reddish brown, and as they advanced a constant rain of violet chain lightning fell on the undulating country below.

Swiftly the horses were unpacked and put in the yard, but before I could get into the humpy a dead tallowwood 50 yards away was struck. I was scarcely inside when there was a sharp crack, my knees doubled up and I went in a heap; the roof had been struck. Very shaky and sick and frightened I got up, pushed out the shutter and looked out; the horses had been knocked down but showed signs of getting up—horses are more sensitive to lightning than men. It was while looking out that I saw something else; two balls of fire were drifting slowly past the humpy about 15 feet from the ground; they were about the size and shape of a soccer football and were a deep glowing red like the coals of a burning ironbark log; they drifted idly this way and that and it was the very uncertainty of their purpose which made them so terrifying. A flash of chain lightning occupies but the merest fraction of a second and if you see it you know that it has missed you, but there is something indescribably horrible about ball lightning; it can hover about you for a minute, drifting lightly as thistledown yet being potent as a ton of dynamite.

This was but the beginning of a bombardment; for nearly an hour incessant waves of red and violet lightning danced through the cracks of the old humpy to the accompaniment of high-pitched, whining crashes which often overlapped each other like machine-gun fire; sometimes my spine would contract, and a numbness go through me from induction of some close flash. At times

I looked out; the horses were weathering it all right; always there were fireballs drifting; at times they exploded and the red light which flooded the humpy brought with it a wave of heat. Like all good things or bad, the storm passed. Tom had been caught on the high Cainbable ridge; he secured his mare and ran down the eastern side of the spur, where he found dubious shelter under the side of a box log. He received a bad shaking from shock and at times had felt the suffocating heat of bursting fireballs."

Case 4: Santa Maria 1902

Eruption of Santa Maria Volcano, Guatemala 1902.
  Eruption sequence of events

- 24 October: 5:00 pm: At San Felipe a horrible sound was heard, similar to the roar of a waterfall, for 5 minutes, coming from the volcano; but the mist surrounding the volcano did not allow any direct observation of what was happening.
- 24 October: 6:00 pm: Cinders and ashes started falling over Quetzaltenango
- 24 October: 7:00 pm: Witnesses recall seeing lightning and a strong fiery red light coming from the volcano, and noise similar to that of an industrial furnace.
- 24 October: 8:00 pm: From San Felipe one could see a giant plume of black ash with numerous fierce twisters crossed by thousands of lightning bolts and curved lines of red light. All the area surrounding the volcano kept shaking and large explosions could be heard as far as 160 km away; strong winds carried ash and debris as far as 800 km (500 mi) away, or even more; a part of the cloud hovered on the north side of the cone for days, and a pitch-black darkness ensued.
- 25 October: 1:00 am: The eruption violence increased and large rocks from the volcano started falling as far as 14 km away, destroying towns and farm houses.
- 26 October: 12:00 am: The volcano calmed down.
- 26 October: 3:00 pm: Another eruption, but this time it was a white plume that came out, which was likely composed of water vapor.

From: Anon (1905) Elektrische Erscheinungen bei den Vulkanausbrüchen in Mittelamerika, Meteorologische Zeitschrift März 1905, 139–140.

San Cristóbal Cucho, Guatemala, about 30 km to the north-west of the volcano.

We take the following parts from the report of Mr. Karl List, an engineer:

"On 25 October, in Cucho (on the lee side of the volcano) strong electrical discharges during the eruption. The electricity left the clothes, the body of the people and the houses. Ball lightning everywhere, exploding with a hollow sound without doing damage (Cucho downwind of the volcano!).

Rain from thunderstorm clouds fell copiously from 250 to 1600 meters, little below that not at all above 1500 meters (??). Above everything was covered in dry volcanic sand and ashes. The volcanoes looked like they were covered in snow far down. With northerly wind, everything is engulfed in thick dust, or dust or rain clouds. In the evening the most wonderful, gorgeous sunsets; in the north and east deeply black rain clouds, saturated with volcanic dust, which are dissolved after sunset by the heat radiated from the earth. One can see the remaining, very thin rising ash clouds, after 7 pm shimmering in rose-red in the infinite heights.

At 8 pm the sky is usually completely clear, and no lighting effects can be seen. Curious cirrus clouds occur very often. They are very high. One can practically see how the ash particles are frozen into the ice fog; they wrap and cover the cirrus like a gray veil.

The thunderstorms' characteristics were curious, because of the breaking of the thunder in the ash particles of the clouds, a never-ending, hollow rolling sound without thunderclaps. The lightning were broad flashes, rarely ball lightning, no zig-zag lightning at all; in short, the thunderstorms were completely different from the usual tropical thunderstorms.

Excerpts from a letter from Mr. Karl List on 25 November 1902."

Case 5: Amiens 1884, Brand Case 55

From (Decharme, 1884).

Thunderstorm and lightning in spherical fragments observed in Amiens, on 24 February 1884.

On Saturday, 24 February, after a relatively warm day (10 degrees centigrade), after a sequence of various different gusts of wind (from the southwest), rain, graupel, and even hail showers, a thunderstorm discharged. At 7.45 pm an exceptionally strong lightning stroke illuminated the whole town, followed immediately (within 2 seconds) by a terrible thunderclap which shook the window panes of all the houses and frightened many people, because of its sudden appearance as well as its vehemence.

Even though lightning and thunder occurred only once in this thunderstorm, the lightning hit simultaneously in several places (which were at a considerable distance from each other) in the same form, which means in globular fragments, which appear to have been diminutive versions of the very rare ball lightning.

1. The lightning hit the theater first, during a performance. It broke through a window pane leading to a small courtyard in the east, facing the law courts (Figs. 1.1 and 1.2); from there it moved towards the backstage where several actors were present, passing one of them very closely, but this person did not feel any jolting (although his trousers were slightly singed above the knee). It was seen passing for more than a second in the form of a small bluish sphere of fire of 2–3 cm diameter, provoking a very small explosion like the lighting of a match. It moved over the side, vanishing below the stage, where it was immediately checked that it had done no damage. Everybody had escaped, fortunately only with a shock.

   The almost circular or elliptical hole which the lightning made in the window pane, the only trace it left in its course, was 3 cm long and 2.5 cm wide; all its edges were broken and it showed no trace of dissolution (melting). Note that the opening was only 6 m above the ground and was sheltered by the theater building; furthermore, near the side of the window which was broken by the lightning, there was a long metal pipe which was connected to the ground by the water which was running abundantly during the thunderstorm.

2. At the same moment the lightning hit a private home (of H. Gossart), situated about 200 m south of the theater. It entered through an open window facing west, and appeared to two persons in the form of a small sphere of fire as large as a nut. When it reached the area above a table, where a young man was writing, it exploded with a bang close to his head. This apparition had not lasted more than 2 seconds, when a terrible thunder clap was heard, and they realized that they had been visited by lightning in a rare form. Even though the shock was great, there had been no accident or damage.

3. At the same moment the lightning hit the chimney of the town hall 400 m away from the theater (not far from one of the lightning rods of this building), namely in the "Bureau Central de Police", where two officers saw a small electric glow and heard a small explosion, as if a firecracker had exploded. This was also a tiny form of ball lightning.

The electrical alarm systems went off, radiating from the main building to the four corners of the city. (At the telegraph office the service was not interrupted, since the lines were below ground in the part of the city affected by the thunderstorm.)

4. In the house of Mr. Camand (30, rue Saint-Denis, 150 m away from the theater, behind the law courts, which was equipped with lighting rods), a maid who was in front of the open kitchen door leading to the court when the lightning struck, saw a white-bluish flame, with a shape that could not be described clearly, and which, coming from the south-west, fell on a drainage pipe next to the door. The women saw the fireball, which appeared not larger than an egg, fall near her feet. When it reached the drainage pipe, the flame made a noise which was similar to a gun shot, or better, similar to a firecracker. The frightened woman ran into the kitchen and had just enough time to sit down when she heard the terrible thunderclap. Note that at this moment a very large amount of water was flowing out of the cast-iron pipe onto the road. Here too, the lightning produced only fear and no damage and left neither a trace of smoke nor a smell.

5. In the house of H. Coquel (44, rue des Sergent, 270 m away from the theater in the direction of the Jardin des Plantes), lightning came down the chimney into the kitchen of H. Braut and raised the cover of the stove, where a fire was burning, like a valve. A fireball ascended from the stove. It was observed and precisely described by Mme. Braut. The flame was initially very voluminous, but when it was 2 m away from the stove in the middle of the kitchen it shrank to the size of an orange or an egg. It then exploded with a sound like a gunshot and disappeared without trace of smoke or smell. This all happened in less than 2 seconds between the lightning and the thunder.

   Below the kitchen there were gas pipes, which had recently been taken out of service and which consequently contained a bit of gas. When another person went down to this place to see if the lighting had done any damage, she saw a small flame at the outer end of one of these pipes, which only could have been produced by the lightning. It should be noted that the kitchen was connected to this basement floor by a metal pipe. All this indicates that after leaving the kitchen the lightning went there over the pipes, before it reached the wet ground at this point.

6. When the lightning struck, an employee of H. Gammad was eating in the company of several people in a restaurant near the railway line and the station on the corner of rue Jules-Barny (590 m from the theater). Shortly before the thunder, he saw a blue flame running over the table at great speed. His company didn't have the time to see it. It dissolved without

noise and left no trace. It is not known how it entered the house; maybe through the chimney.

7. Finally, the lightning hit the house of H. Guilbert, which is on the Boulevard du Jardin-des-Plantes, 800 m from the theater. The shape of the lightning was the same as in the house of H. Gossart (a nut shape). Everything indicates that it reached there via a telephone cable which was extended and stretched by several centimeters. Moreover, the bell, to which it led, was damaged. Leaving this wire in the kitchen, the lightning made an explosion at least as loud as a gunshot. After this noise was heard, a blue flame (color of a punch?) of the size of a nut moved from the kitchen into the neighboring room. It almost circled the table, where several people were eating, brushed the host whose hand was numb for several moments and then vanished without trace or smell. Everything happened in less than 2 seconds and only then the strong thunderclap was heard which explained the event. The people were very much disturbed, and a man servant suffered a shock.

Even though there was only one lightning strike during this thunderstorm, it came down in globular form in seven different places. The two most distant ones were at a distance of about 1360 m.

One can conclude from the above facts that the lightning had especially unusual characteristics, because it separated into spherical fragments which were harmless because of their small size.

An experienced observer told us that, at the time of the lightning, he saw from his open window a large luminous mass of exceptional brightness, which blinded him for a few moments. This mass dissolved into an indistinct fog which distributed itself over various points of the city, and especially in the direction of the places mentioned above.

We are dealing here with a new type or sub-type of lightning, which is a smaller version of the not yet completely understood phenomenon of ball lightning.

Case 5: Tissey 1912 Brand Case 184

2 December 1912. (3 pm?). Tissey (Yonne), Comptes Rendus 155, 1567.

On 2 December 1912, at 1 o'clock the sky over a village 8 km from Tonnerre (Yonne) to the north of Tissey became overcast after a rainy morning, which was followed by a long period of clear sky. Soon a storm with rolling thunder and lightning broke loose over Tissey, a very dense rainfall drenched the soil and then the storm seemed to calm down. It was about half an hour after the last thunder had been heard; the wind had died down, the sky was uniformly covered with cloud, the rain fell steadily; suddenly a heavy thunder clap was heard, immediately followed by a most powerful discharge which extended over a distance of l km over the entire village. Over most parts of Tissey and simultaneously at all points, bursts of fire spouted out of the ground. Above a water reservoir, which was supplied by a subterranean water layer, the discharge took on the form of a sphere three times the size of a head and was completely separated from the water surface. At the same moment a laborer, who was standing in a shed 100 m from the water reservoir, saw a head-sized fireball pass on the street at a distance above the ground corresponding to the height of a man of average height. The appearance of the ball was, so he said, separated from the observation of the ordinary lightning which he saw an infinitely short but noticeable time after the fireball. Outside the village the discharge had melted a steel wire, which was stretched over a hay stack and anchored by two large stones, over a distance of about 15 m and caused a fire to break out, which, however, was quickly extinguished by the rain. The ground was uprooted over a length of 2–3 m and to a depth of 5 cm at the points where the wire ends fell, dragged along with the stones. The discharge was followed by a downpour accompanied by hail and snow. The above facts indicate that the appearance of the ball lightning was not due to circumstances whereby the position of the observer had caused him to see zigzag lightning straight on. A comparison of the present case with that observed at Strasbourg on the Rhine in 1869 and described by Ch. Hugueny points to the similarity of the conditions and the presence of subterranean water below the sites where ball lightning was observed.

Case 6: Arnsberg 1868 Brand Case 30

30 August 1868. During the day. Near Arnsberg. Das Wetter 6, 68, 1889.

In August (?) of the year 1868 I was engaged in the construction of a tunnel on the Upper Ruhr Valley railroad in the vicinity of the town of Arnsberg. One day during a thunderstorm we took shelter under the roof of the nearby blacksmith and wheelwright shop; the shelter provided by the roof against the down-pouring rain was not too good, since the roof had leaks at various places. The sun was already shining again for some time and only a few, fairly transparent clouds could be seen in the sky. We had already forgotten the storm and were even less occupied with the thought of another thunderstorm. The four of us were busy on the trestle and were just about to lift a stone plate approximately 80 cm square onto the wall; we were standing in a circle, suspecting nothing, when suddenly lightning flashed and a round, yellowish, transparent ball of about 20 cm diameter appeared in our midst at approximately 90 cm above the stone we were about to lift onto the wall: the ball oscillated continually about 4 cm up and down above the plate. At the center of this ball there was a blue flame, which was pear-shaped with the tapered portion slanted downward; it had a length of 4 cm. This flame rapidly rotated around a vertical circle of about 7 cm diameter inside the large ball. Anyone can imagine how startled we must have been. My eyes were fixed upon the frightful intruder and my only thoughts were how to get rid of it. However, I shall not waste time elaborating on this. After about 3–4 seconds there followed a loud bang, such as I had never heard before, and the ball vanished, none of us knowing where it had gone. However, we breathed more freely now, and I felt as if the full load of heavy stone which lay before our feet had been taken off my shoulders. When we recovered from the shock and lifted the stone onto the wall our limbs were almost paralyzed.

The wheelwright, who was working on the roof about 10 m from the site of the event, did not notice the ball, but upon hearing the detonation slipped either because of fright or the shock wave, and fell off the roof without suffering any injury. Immediately thereafter we were informed that lightning had struck a quarry about 100 m away.

The editor in chief of the newspaper "Das Wetter" added: considering the rarity of cases whereby ball lightning is clearly observed, it seems appropriate to publish earlier accounts of this phenomenon, provided that they were given by trustworthy persons. The above description originated from chief mason Emde of Brilon and amply fulfills the said prerequisite concerning the meteorological observations made. The same view was put forward by the newspaper "Sauerländer Anzeiger."

Case 7: Loeb's Account

From (Humphreys, 1936).

"The following quotation from a letter by another eminent physicist, Leonard B. Loeb of the University of California, Berkeley, is also interesting. "I am particularly anxious," he assures us, "to record this experience in view of the fact that it occurred to me when I was a young child of some eight to ten years and, although I have never been imaginative and given to storytelling, I was laughed at for my statement. I can, however, remember it as clearly today as when it occurred; incidentally at that age I had not heard of the phenomenon before, so it is not a figment of imagination. I do not believe anyone else saw it, at least no one but my brother was near, and he was much too young.

It was during a summer thundershower in Springfield, Massachusetts, and must have been around 1898 or 1899. It was an afternoon thundershower, occurring, as near as I can remember, between 3 and 5 o' clock, probably at about four. The phenomenon occurred at the beginning of the storm, that is, as the main thundercloud was approaching; it was already fairly dark. I was indoors on account of the impending shower and was observing it from the front window of my grandfather's house. It occurred coincident with a striking of the lightning on the cornice or roof of a house across the street and one or two doors up. It preceded the thunderclap and the flash. As I looked out of the window, I noticed a ball of what I would now describe as the color of active nitrogen or possibly slightly darker, as it seemed to me, descending from somewhat the direction of the neighbor's house in a light graceful curve. Its diameter appeared to be about double that of the toy balloons which one sees and its motion through the air was quite analogous to the motion of the type of air-inflated balloons which are used so frequently in modern dinner parties. It had a translatory motion in my direction and seemed to descend down an inclined plane from the approximate location mentioned. It appeared to strike on the lawn, bounced slightly once and then disappeared. Its disappearance was followed, better accompanied, by a tremendous clap of thunder and flash of lightning which appeared simultaneously. This was the flash which struck the cornice. There was no visible after-effect and its outline, so far as I remember, was more or less indistinct, although it was quite spherical in shape. In the question of the sharpness of outline I am no longer definitely certain. As regards the direction of the wind, I am inclined to believe that it might have followed the direction of the wind, although at that particular time the lull between the up-draft and the thunderstorm wind was on. It has been my impression, as I have thought of it in later years, that the phenomenon was caused by some type of intense glow discharge caused by the effect of the field of the approaching cloud on the cornice or some projecting angle of the building. It has been my impression in thinking of it that the so-called ball was internally in rapid rotation of some sort or that there was a vortex which gave it its shape. The color was definitely that of active nitrogen, as I have since seen active nitrogen in the laboratory."

Case 8: From Humphreys (1936)

It was described by Dr. Joseph S. Ames, long-time professor of physics at the Johns Hopkins University. In his letter of 19 June 1924, quoted with his explicit permission, he says: "Mrs. Arnes was standing on a rug during a thunderstorm with her hand at her waist, one finger more or less extended. I was about 5 feet away and noticed the air between her finger and the floor was quivering so that it looked just like the hot air over a field. I noticed something rise slowly from the floor up towards her finger and then there was for an instant a small oblong fireball about the size of a pecan attached to her finger. It was not very bright and appeared to shine through a haze. There came a flood of lightning outside and the fireball disappeared."

Case 9: Neustadt Multiple Ball Lightning Case (Wittmann)

NEUSTADT MULTIPLE BALL LIGHTNING CASE.
Originally published in: Keul, A.G. (ed.), *Progress in Ball Lightning Research*, Proc. interdisciplinary congress Vizotum held at Salzburg, Austria, 20–22 September 1993, pp. 110–114, Salzburg (1993).
Axel D. Wittmann, University Observatory Göttingen, Germany.

An extraordinary case of multiple ball lightning observed in 1951 by the author and several other witnesses at Neustadt (near Coburg/Germany) has been reported previously. An eyewitness account written immediately after the event by one of us (Mr. C.W. Förster) has recently been found among the papers of the late Mr. Förster. His report not only provides us with an independent description of the event (in addition to that of the present author), but also includes the previously lost information about date and time of occurrence. The present paper gives a complete report based on these two accounts and some additional on-site research.

1. The Event
In an earlier paper the author has published an eyewitness account of a quite unusual case of ball lightning which occurred during a thunderstorm with heavy rain at Neustadt (near Coburg/Germany). In brief, a bright but not extremely blinding luminous ball of spherical shape, colored orange to yellow, with a diameter of 60 ± 30 cm was observed above a lime-tree at a height 18 ± 2 m above the ground. When first seen, the ball moved vertically downwards with a velocity of 3.5 ± 0.5 m/s until it reached the tree-top at a height of 10 ± 1 m above the ground. When it made contact with the tree, the plasma ball instantly disintegrated into about 12 ± 3 smaller balls: these were of equal

size (each having a diameter of 13 ± 2 cm) and of approximately the same color and brightness as the primary ball. The fragments (whose total volume may have been less than that of the parent body) fell slowly towards the ground, moving along the outer contour (tip of branches) of the tree; they fell vertically during the last few meters in the absence of branches. All of this occurred without much noise (in particular, no bang was heard at the moment of split-up). On reaching the ground (a pathway and a roadway) about 4 ± 1 second after split-up, the luminous balls instantly disappeared without noise (beyond the level caused by the rain). Where the lightning balls had fallen on the wet asphalt, circular impact marks were still visible when the author inspected the site after the rain had stopped about 30 minutes after the event. These resembled circular oil spills with a diameter of 12–15 cm, and they showed the interference colors of thin layers. From the energy needed to evaporate the water layer and to melt the asphalt surface (assuming it contained the standard mixture of basalt and *B80* bitumen), one may roughly calculate an energy density of ~2×10^7 J/m^3, or a total initial energy content of about 0.6 kWh.

Case 10: Hahnenklee Brand Case 192

"On June 22, 1914, ball lightning penetrated the glass-framed veranda of the Schwenzel Hotel in Hahnenklee. Mr. Kuhlgatz, the local school inspector in Kiel, who had observed the phenomenon from very close by and added further details concerning the local conditions after making a second trip in September to the above-mentioned hotel, was kind enough to communicate the following information. Mr. Kuhlgatz was sitting between 6 and 6.30 o'clock in the evening on the above-mentioned veranda; his chair was positioned such that the long glass partition of the veranda was to his left, i.e., his line of sight was directed toward the fairly distant side partition which closed off the veranda. Two other persons were sitting at his table, and at another table in front of him there was a second group of guests. While all the windows were shut, and the rain was pouring down outside, Mr. Kuhlgatz suddenly saw ahead to his left a fireball about 10–15 cm in diameter penetrate via the upper window into the veranda and move around with moderate velocity. After a few moments there followed a cannon-like discharge. The second group of guests, who were sitting directly below the path of the ball, jumped up from their seats. In the guestroom adjoining the veranda there was a washbasin next to the bar which was connected to a water pipe about 1 m away. According to another guest, who was standing next to the bar, the path of the fireball was observed in the direct vicinity in or inside this basin. The fireball left no traces whatsoever on the walls of the veranda or in the guestroom. On the other hand, the telephone, bell, and electrical lighting wires of the house were heavily affected. They were in part destroyed and, in some place, fused. All the electrical lamps went out and the bell rang incessantly. No ozone- or sulfur-like smell was perceived. Several minutes later, an intense, strongly twisting zigzag lightning flash was observed on the side of the windows. The question is how the ball lightning entered the veranda and the adjoining room. According to direct observation, the ball had penetrated through the window partition of the veranda, which was everywhere closed, and passed from here into the guestroom, which was likewise closed off by a wood-paneled partition with inlaid windows. In neither of the two partitions were there any openings or perceptible slits between the window panes. The telephone and electrical-lighting wires are outside at a point on the house wall, which projects considerably above the veranda, about 1 1/2 m above the veranda root which is covered with roofing felt. This point is about 6 m away from the point at which the fireball entered. All the wires in the veranda come from within the house and in part run along the wall and in part along the ceiling; there were no wires along the window partition of the veranda. Thus, the ball lightning had passed through a tightly closed partition consisting of wood panels and windows without damaging the latter in any way. If we wish to understand the causes behind this extremely remarkable event, then it seems that we are left only with the following explanation. The wires running through the guestroom and the veranda acquire a very high charge from the overhead lines leading to the house. This charge gave rise to the creation of spherical, luminous, and exploding formations on the veranda as well as in the guestroom. Qualitatively, this can be correlated with other experiences. Here we will merely assume that quantitatively the balls which jumped from the electrical wire must have possessed an uncommon size and explosive power. Of course, in this interpretation we must assume the possibility of a slight hallucination of the observers, insofar as the path of the ball did not start on the window side. I believe the second interpretation is the more probable one."

Case 11: Sagan Page 181 (Sagan, 2004)

"At Needs Laboratory at Wright Patterson Air Force Base, in 1954, I generated fireballs regularly, and what they did varied. Once I operated a sputtering regulator (jig) to electrode solar batteries with a 1000-volt second-anode discharge at 40 mA. Suddenly the equipment sneezed, and from the 3/4-inch pump exhaust came a 5-inch purple-pink fireball with sharply-defined edges that pulsated, changing shape from oblate to round to ellipsoidal, but always changing back like a rubber balloon full of slow-motion water. Standing two feet from it, I got a good look. It was as bright as a 20-W bulb but never changed size or color. Airborne, it glided out the door into the hallway, turned left and floated down the hall, westward.

Finally, it floated through a glass windowpane like it was not even there. But we never saw where they went after that, perhaps into some trees. There was always a smell of ozone, but no smoke. This occurred often, and whenever we heard the equipment start sneezing, we knew a fireball was coming out, and everybody ducked for cover. We even nicknamed it "The Sneezer." After five times, the fireballs came more frequently. Finally, one fireball did not go out the door. Instead it struck a large instrument panel and exploded, burning out the panel. Sparks jumped out and smoke poured out. After that we were ordered not to use the equipment, which sat idle in a corner."

Case 12: Turner (1998)

"An observation recently reported to the author (by Mrs. S. Orser of Huntingtown, Maryland) is particularly interesting as it has a number of unusual features some of which have a bearing on points which have just been discussed. The event occurred at Mrs. Orser's home during late June or early July of 1989. The atmospheric conditions were very typical of those during which balls normally form near the earth: a severe thunderstorm was in progress nearby and it was extremely humid. No window in the house was open because of the strong wind outside. Although she cannot now be certain whether the air conditioner was operating, it was her normal practice to switch it off during a thunderstorm so that in all probability it had been switched off. The clear association in her mind between the very humid conditions during the storm and the observation supports this belief. The ball "suddenly appeared" indoors directly in front of the observer at a range of just over 3 m and quite stationary. Although the observer was sitting almost directly facing it, its formation stages were not noticed. It appeared almost directly underneath a large skylight made of Plexiglas (perspex) and it remained, so far as Mrs. Orser could tell, absolutely motionless for the whole minute it was present. Thus, she had plenty of time to observe it and listen to its crackling sound. It then disappeared without additional noise and, about 45 minutes later, a second ball, somewhat smaller in size, appeared for a few seconds underneath a different skylight. Mrs. Orser obviously could not observe this one so carefully and the detailed descriptions apply to the first ball. The ball was unusual both for its long life (particularly for a ball indoors) and its exceptional positional stability. For a ball formed indoors it was also unusually large, being almost a meter and a half in diameter. The second ball was about a meter in diameter. In appearance, the balls were bright bluish white with the bright white more towards the center (at least for the better observed one) and the blue more towards the edge. As an indication of brightness, Mrs. Orser noted that the ball was not painful to look at, a bit like white paper illuminated by a bright but diffused light source. It was, however, translucent to transparent. The translucent portions were where small electrical discharges seemed to be continually forming. The crackling noise was presumed to be associated with these "worm like discharges". The lightning ball was clearly associated with a strong electric field which caused the family cat's fur to stand on end. The cat did not seem concerned until Mrs. Orser attempted to pet her. A sharp discharge then took place between the two when Mrs. Orser's hand was about 38 cm from the cat. After this shock the cat did seem disturbed. Mrs. Orser was not frightened but suffered of the symptoms which have been reported by others who have experienced close proximity to powerful ball lightning or ordinary lightning. As might be expected from the high field, a strong smell of ozone persisted after the ball expired. A comment was made that, though she was fairly sure the gas was mostly ozone, it did not smell quite like the gas responsible for southern California smog. Mrs. Orser could not be specific about the length of the glowing "worms" because they were continually forming and breaking up. Their observation looks suspiciously like evidence for dust particles aggregating and separating. Since pets share the house, sizable dust concentrations are to be considered normal. With respect to inertial forces, the most significant aspect seemed to be the fact that the ball stood absolutely motionless for a minute. The present author was particularly intrigued by this observation as it contrasts so strongly with what is normal. As a consequence, and bearing in mind Dmitriev's observation, he later contacted Mrs. Orser for a second time to ask her if there was, in the room, any large conducting object which had sharp edges and circular symmetry. Her immediate reply was that there was just such an object more or less underneath the position of the ball. This was a heavily engraved brass table (bought in Morocco) a meter and a quarter in diameter. Clearly the ball's positional stability could represent the same guidance phenomenon as Dmitriev's but with circular, instead of linear symmetry."

Case 13: Rakov/Uman (2003) 20.2. Outdoors in Australia

Near Murray's dairy, a property on the Queanbeyan road on the then out-skirts of Canberra, and not far from what is now the industrial area of Fyshwick, I was riding along the right-hand side of the road, just off the pav-ing to prevent a possible slip and fall on the wet surface, and about 20–25 yards (or meters) ahead, on the left-hand side and also off the bitumen, a farm employee was leading a shorthorn bull. There was very little, if any rain at that time. I can clearly recall that there was one of those periods of "quietness" that sometimes precedes a downpour. Just as I drew level with the bull there was one very loud bang or explosion and immediately down the white traffic line in the center of the road appeared the fireball. It seemed to be about 6 or 8 inches (15–20 cm) off the ground, was about the size of a basketball, like very golden butter in color, and had the appearance of being "spun" or "fuzzy", like silk threads or wool, as distinct from a "molten" liquid look. It did not spar-kle—just a ball of fuzz. It came straight down say three of the white marker lines. These have about equal distance of unmarked road between them, so I should say it traveled about 18–20 ft (6 m). Then it simply disappeared. It did not break apart. There was no further noise like an explosion. It was there one moment and not there the next. The whole thing would have been over in probably 2 or 3 seconds, before the horse had time to be startled. The young men leading the bull cried out—in pure Australian—"What the bloody hell was that?" As there was another downpour shortly, we did not stop to discuss it.

Case 14: Cavendish Laboratory, Extreme Weather: Forty Years of Tornado and Storm Research

Editor: Robert K. Doe, Wiley 2016, p. 226.

Professor Sir Brian Pippard FRS (1920–2008) reported a particularly well attested event observed at the Cavendish Laboratory of the University of Cambridge just after 4 pm. during a violent thunderstorm on 3 August 1982. The storm was so severe that staff on the upper floor of the two-story building considered going downstairs. Lightning struck the laboratory and its vicinity several times, although there was no structural damage. Immediately follow-ing one flash near the Bragg Building several observers reported that they saw at least one luminous ball. A physicist seated with his back to a window on the ground floor of the Mott building saw his |room briefly illuminated as if by a very bright object moving rapidly westwards between the Bragg and Mott buildings. It would be difficult to explain his observation as a positive afterimage. A second observer on the first floor saw the space between the buildings 'filled with a luminous haze' at least to first-floor level, and he interpreted this as sheet lightning. However, on looking to the west, he observed a blue-white light of approximately the apparent size of the moon which appeared motionless about 10°–15° above the horizon and was in sight for 3–4 seconds.

Another observer in the same room may have seen the same phenomenon just before then, as she had the impression that it was receding while expanding from its original size which was about that of a grapefruit. The data are consistent with an approximate distance from the observers of 12 m. Three other people made a further report of a very bright, blue-white ball moving above the ground to the west. They also said that it subtended about the same angle as the moon. It was in sight for about 4–5 seconds before it vanished.

A closer encounter was experienced by an administrative assistant in a duplicating room on the ground floor. She was closing a window when she was startled by a noise that suggested the window had been knocked in. A bright, spinning, sparkling object of pyrotechnic appearance entered past her head, rebounded from a copying machine and departed as it had arrived. She said, 'It came in through the window, spinning, rolling, throwing out all sorts of sparks like a Catherine wheel. I was terrified'. The window was undamaged. Another person in the same room was convinced that something had entered the room.

Case 15: 2017 in Paignton, Devon

Report online:
https://www.devonlive.com/news/devon-news/pensioners-horror-ultra-rare-ball-457467

"Then a bright blue ball came through my window, moved across in front of me, then went out through the glass patio door. It took about a second and a half.

It had an orange tail to it. I thought I was seeing things. I thought I was going round the bend, but I know what I saw. The bang was absolutely unbelievable, and the ball was like something out of a sci-fi film." Mr. Dodd[2] said

---

[2] Unfortunately, Mr. Dodd passed away in early 2018. I am very grateful that he gave such a detailed report of this unusual event.

his wife Margaret was in the kitchen of the apartment at the time and did not see the phenomenon.

Neighbor Wendy Holmes did see the ball lightning as it crossed the car park of the Bosun's Point luxury retirement apartments development where she and Mr. Dodd live. She said: "It came with a huge clap of thunder, and it has done quite a lot of damage to the electrics. The electrics in the lifts have been burned out, and all the Sky TV boxes have been damaged. "It was like a massive blue ball followed by a tail. I saw it going across the car park. It came with such a horrendous explosion. I rushed out onto the balcony to see what had happened."

Personal information from Mr. Dodd in an email:

"It was a ball approx. 30/40 cm diameter. I was 1.5 meters from it. There was no smell, but the thunder noise was more or less at the same time and was VERY loud. The ball was very intense bright blue and was very sharp and you could not see through it. It came in a closed window and out through the glass of the closed patio doors, both window and doors are double glazed. The ball moved from left to right approximately 6 meters and about 1.5 meters in front of me at about 1.5 meters high.

I did not see the ball outside, but Wendy my neighbor did and also it turns out another lady saw it traverse the car park about 30/40 meters and this was after it left me through the glass.

Sorry there are no cameras. The lightning struck the block of flats next to me at the roof level. It destroyed the communal large sky dish and all sky cables to the 7 flats and also destroyed the block's lift control panel plus damage to ceilings and some lights."

The lightning location network EUKLID reported a negative CG strike on the neighboring house (Date and Time: 2017-09-08 13:24:22, maximum current: −15.7 kA).

Case 16: Rakov/Uman (2003), 20.2.4. Indoors in Nebraska

I was standing in the kitchen of my home in Omaha, Nebraska, while a terrible thunderstorm was in progress. A sharp cracking noise caused me to look toward a window pane to my left. Then I saw a round, iridescent (mostly blue) object, baseball size, coming toward me. It curved over my head and went through the isinglass (mica) door of the kitchen range, striking the back of the oven and spattering into brilliant streamers. There was no sound and no effect on me except a tingle as it passed over my hair. Later examination showed a tiny hole with scorched edges in the window pane and isinglass, and scorch marks on the back of the oven.

Case 17: Rakov/Uman (2003), 20.2.5. Indoors in Virginia

I was laying in bed after a thunderstorm, happy that the storm was over. Suddenly, I could see a bluish-colored ball, nearly perfectly round, about the size of a softball or slightly larger, floating outside my bedroom window, perhaps 20 feet above the ground. I could see it through the plastic mini-blinds covering the window, which were closed but not tightly closed (as we usually keep them such that I can frequently see the moon from my bed). After a few seconds, the ball lightning floated into my bedroom through the window. It came through the glass window and the plastic mini-blinds as though the window and blinds were not even there. There was no sound of it hitting the window and no change in direction or the shape of the object as it came through the glass and plastic blinds. It continued floating, quite slowly (slower than soap bubbles blown with a bubble wand) in my direction. After a few seconds, it was about two feet from my head. It 'poofed' - not a loud explosion but a definite 'poof', and then I smelled something, as though something had burned.

I'm sure about the burning smell. Had I not just seen that thing explode and been sure that it was the cause of the smell, I would have gotten up and checked the furnace and other things for a source of a possible fire. The ball of lightning had totally disappeared with that poof. The explosion, however, was considerably softer in its sound than the popping of a balloon. It never occurred to me at the time to check the window for damage, because the ball lightning came through the window as though nothing at all was in its path and without making any sound whatsoever at the point. Checking recently, I saw no damage. The window was a storm window, with one pane of glass, some air space, plus another pane of glass on the outside. The frame of the window was a combination of wood and aluminum. There are no power lines near our house (for at least 1/2 mile) as all power, phone, and cable lines are underground.

Case 18: Rakov/Uman (2003) 20.2.11.
From a Radio Transmitter Operator

Back in the year 1950, midsummer, and in a place called Degendorf nach Brannenburg, Bavaria, I was stationed there as a US army soldier. I was a radio operator with the third Battalion sixth AC Regiment. This particular day, it was about 2 pm. mid-afternoon, and the sky was clear and sunny. I was on radio duty sitting in an enclosed housing on the back of an army 2 1/2-ton truck. The power supply was a hookup from a nearby building (220 V). Normal mobile power was from a towed generator trailer. The interior comprised one wall counter unit, with two receivers and a counter top with mounted telegrapher's key, a long bench or box seat running lengthwise with the truck bed, storage units behind the seating, and at the far end interior a radio transmitter approx. 30 inches × 25 inches (75 cm × 63 cm) and 36 inches (1 m) high. As the truck was in "fixed location" status, we were using a long wire antenna strung between two 2-story building roofs (approx. 200 ft length). The lead-in wire ran from its (the antenna's) center, down and into the truck, and fastened to the antenna post connector on the back of the transmitter. The antenna wire was skinned back of its insulation the first 1 1/2 inches and threaded thru an aperture of the post and fastened with a set screw.

I was sitting inside with a fellow operator, rear door wide open and playing cards while monitoring radio traffic. Locking the telegraph key down to the "send" position, I turned the transmitter on, leaned over it and (taking) a common wooden lead pencil, I put it close to the antenna coupling until I "drew forth" an arc whereof I then leaned in, with my cigarette to mouth, and lit it from the arcing. As I leaned back to sitting again, up from behind the transmitter floated that "shimmering fireball". Shimmering, pulsing, blue of its fire, it then floated right at us (18 inches (0.5 m) in diameter) (of a truth, to now I can't say for sure whether I had yet shut the transmitter off or not). We two young men that we were, this experience was beyond us, and we fought ourselves to bail out of that truck before that object could touch us. Hitting the ground, we turned to look back, and there was nothing to be seen of it anymore.

In retrospect, I would have to say that the fireball had to originate at the same antenna tie-in as where I had lit the cigarette from. It floated up from that very side where the antenna post lead was connected. And then bee-lined straight for the open doorway that we were in front of."

Case 19: Jennison

"[…] I was seated near the front of the passenger cabin of an all-metal airliner on a late-night flight from New York to Washington. The aircraft encountered an electrical storm during which it was enveloped in a sudden bright and loud electrical discharge […]. Some seconds after this a glowing sphere a little more than 20 cm in diameter emerged from the pilot's cabin and passed down the aisle of the aircraft approximately 50 cm from me, maintaining the same height and course for the whole distance over which it could be observed. The observation was remarkable for the following reasons.

* The appearance of the phenomenon in an almost totally screened environment;
* the relative velocity of the ball to that of the containing aircraft was 1.50 ± 0.5 m/s, typical of most ground observations;
* the object seemed perfectly symmetrical in all three dimensions and had no polar or toroidal structure;
* it was slightly limb darkened having an almost solid appearance and indicating that it was optically thick;
* the object did not seem to radiate heat;
* the optical output could be assessed as approximately 5–10 W and its color was blue-white;
* the diameter was 22 ± 2 cm, assessed by eye relative to the surroundings;
* the height above the floor was approximately 75 cm;
* the course was straight down the whole central aisle of the aircraft;
* the object seemed to be in perfect equilibrium;
* the symmetry of the object was such that it was not possible to assess whether or not it was spinning […]" (Jennison, R. C. 1969)

Case 20: "Tub of Water Case" (To the Editor of the "Daily Mail")

Sir, during a thunderstorm I saw a large, red hot ball come down from the sky. It struck our house, cut the telephone wires, burnt the window frame, and then buried itself in a tub of water which was underneath. The water boiled for some minutes afterwards, but when it was cool enough for me to search, I could find nothing in it.

Dorstone, Hereford W. Morris.

Case 21: From McNally (1966), Observation No. 46

"The above occurred once in Iowa, later in western Nebraska. Both involved the old 'crank type' country telephone. During electrical storms my father taught my seven brothers and me to always aim the phone mouthpiece toward the ceiling to minimize the chances of our being struck by a, quote, 'lightning ball'. At the time I was both too frightened and too small to observe any scientific aspects of the phenomenon, however, I recall seeing the red ball flash from phone to ceiling similar to a Roman candle discharge, except of much shorter duration."

Case 22: From McNally (1966), Observation No. 49

"Date was late summer 1939 or 1940, place Palestine, Ill. There was a thunderstorm in progress. I do not recall definitely whether it was raining before the event, but I believe it had not yet started raining. Later it rained rather hard. My memory is that it was simply 'working up to it' with thunder and lightning. My little sister had gone into a bedroom to use the phone when I heard her cry out and call my name. I ran into the room in time to see a ball of fire settling to the floor much as an air-filled balloon would. It struck the floor and rebounded slightly, again, about 1ike a balloon. There was a rotary motion of the object. It rolled and bounced (damped bounce like a light ball) across the floor to the leg of an iron bed. There was a blue flash, a rather loud spark discharge sound, a little vapor or smoke visible, and the thing was gone. I seem to recall that the white enamel on the bed leg was burned a little, but nothing dramatic. I might say that the sphere seemed to deflect on hitting the floor, that is, made a flat contact spot. The phone stood on a stand near the other end of the bed, and I seem to recall some malfunction of the phone, the particulars of which I forget. My sister told me that she called out when she saw the thing descending rapidly outside the window, and that it came down and through the steel wire screen (!) and bounced on the window sill. I might state that I find myself in the very unscientific shape of depending totally on my memory—I did not write a thing down at the time!"

Case 23: From McNally (1966), Observation No. l69

"I saw the ball lightning while feeding cattle on a farm in middle Tennessee in l946. A bolt of lightning struck an electric fence or close to the electric fence about 600 ft from the end of the fence where I was standing. After the strike, 3 balls of fire shot in rapid succession from the end of the fence and traveled about 200 ft in the air (in a straight line) and hit the ground. The only noise I heard was the thunder and 3 'bangs' when the balls of fire left the fence. The pasture had a dehydrated appearance where the balls hit the earth. I was unable to tell where or what the initial bolt of lightning had struck."

Case 24: From McNally (1966), Observation No. 5

"This was noted during an intense electrical storm. Lightning struck a telephone line and the ball was discharged from the phone. It appeared to be a sphere about 4–6 inches in diameter which 'floated' by a screen door when it disappeared. I was a child at the time and was not interested in detailed observation but was merely frightened."

Case 25: Bead Lightning, Own Observation

This report is on the observation of several bead lightning strokes during a short thunderstorm. The event took place in the evening of 16 May 1994, in a small town in a mountainous region in western parts of Germany (Niederehe, Coordinates E 6.761, N 50.313, elevation 420 m). The thunderstorm moved rapidly from SW to NE, its most active region passing slightly west of me within 15 minutes. When the active region was close, I observed two bead lightning strokes. Later, I saw two additional bead lightning strokes at larger distances and finally a possible fifth one in the far distance. It appeared that all these bead lightning strokes originated from more or less the same region of the thunderstorm cloud, which moved with the southwesterly wind. Each of them had a discontinuous trace where bright regions alternated with shorter dark ones, generating the distinct impression that the trace was broken at irregular intervals. This stroke-dotted appearance could be clearly observed for the two nearer strokes. All the four strokes with a discontinuous trace were strong, more or less straight, cloud-to-ground strokes with little or no branching. None of the other lightning strokes of this thunderstorm had a similar appearance. Attempts to take photos of the bead lightning were made after the second event but failed since they were taken with a camera in open-shutter mode, integrating over the temporal evolution of the lightning. I tried to correlate these observations with information from a lightning detection network BLIDS from Siemens. For this region of Germany, a detection efficiency between 90% and 95% and a location accuracy of 1 km is claimed. All recorded strokes are negative cloud-to-ground strokes except one intra-cloud stroke. Among the recorded strokes there are three with a rather high but not exceptional current (−52,000, −46,000, and −46,000 A). A one-to-one correlation between observations and recorded events is not possible due to the imprecise timing of the observations and to a lesser degree to the rather large error in the location of the observation and the lightning detection, which make unambiguous identification hard. Nevertheless, three candidates can be identified for the first two observed events. All these candidates are normal negative cloud-to-ground strikes with typical current strengths. The bead lightning events reported here differ from other observations in one respect: the luminous regions of the trace were not spherical, but the lightning channel appeared to be broken up into long and narrow segments of varying length. The duration of the visibility of the trace is of course difficult to estimate, but it was almost certainly less than 1 second.

Case 26: Ball Lightning Over Berlin

Report by Wilfried Heil.

We have just observed a very bright luminous feature below a cloud in the morning sky, shortly before a thunderstorm, which slowly evolved into a luminous sphere of about 1 m diameter while hovering at a height of 300 m. The entire process lasted 7–8 minutes, during which time the sphere drifted through the sky at more or less the same altitude and then disappeared into the low hanging clouds.

The event took place over midtown Berlin, on 29 July 2006 at 3:10 am (this morning).

Looking northeast just before dawn on a well-lit street, we saw what looked like a burning plane heading in for a crash or maybe a helicopter with searchlights on. We were concerned as to where it would drop. We had about 80% overcast sky with a cloud level at perhaps 400–500 m. The lights of the object appeared to be about 1 km away.

The luminous feature we saw had developed just below the clouds. It was about 10–15 m in diameter by the time we noticed it, very bright with a steady light output and somewhat irregular in form. At first it looked like a burning airplane which was losing considerable amounts of debris. Within a minute, it changed into what now appeared like a bright fuzzy cloud. There were no discernible features and no sharp edges. The color was at first yellow, then it became a deep reddish orange, rather unlike airplane lights. The light was intense, maybe equivalent to the 10–25 kW of sodium street lighting lamps. The light output appeared to increase continuously for the first 2 minutes, perhaps because the object was approaching.

There was a thunderstorm just forming after a month of tropical heat, with the thermometer around 38 °C in the daytime. We had already received a few drops of rain but had had no observable thunder or lightning yet. The feature appeared out of the clouds without any lightning that we could have noticed. It made no noise.

During the next few minutes, the luminous cloud slowly drifted across the sky with the wind in our direction, at a pace of 10 km/h, walking speed. After an initial increase, it lost its brightness at an exponential rate of about 50% per minute. At the same time the fuzzy halo decreased in size and we gained visibility of a luminous sphere of 1–1.5 m diameter. This sphere was clearly delimited by a deep red outer rim and had a somewhat transparent, also luminous, but blue-greenish-yellow inner region. This inner region appeared to be speckled. It looked like a small balloon or a luminescent bubble in midair.

After watching for several minutes and with the object still hovering in the sky, we started to dash for our apartment (which took just under 5 minutes, timed), in the hope that we might be able to catch it on film and aim a telescope at it. At the time we arrived, the luminous sphere was still hovering between the clouds, almost straight above us. The cloud level had decreased to 250 m, and after a few moments the object disappeared into the clouds.

Immediately thereafter it started to rain. We saw a single cloud-cloud lightning bolt. In other areas of Germany there were fierce thunderstorms during this night which caused considerable damage. The object that we saw first appeared over the sky of Prenzlauer Berg as seen from the street Kastanienallee. We believe that we noticed it immediately after its formation and we were able to track it during its entire development, until it disappeared over Zionskirchplatz about 8 minutes later, almost directly above us and about 1–2 km from the position where we had first seen it.

TIMES: The times were estimated in retrospect. We jogged the same distance again with a stopwatch. The true time of the first appearance will have been between 3:05 and 3:20 am.

SIZE: The sizes were estimated from the perceived distance and the angular width of the object. All distances are difficult to assess in the dark, with only the clouds as a reference. The error might be at most a factor of 2 in either direction.

References

Abbott, B. P. et al. (2017) GW170817: observation of gravitational waves from a binary neutron star inspiral. Physical Review Letters 119:161101

Arago, F. (1838) Sur le tonnerre. Annuare au Roi par le Bureau des Longitudes, Notices Scient.

Arnhoff, G. (1992) Is there yet an explanation of ball lightning? European Transactions on Electrical Power 2:137–142

Bäcker, D. and Boerner, H. and Näther, K. and S. (2007) Multiple Ball Lightning Observations at Neuruppin, Germany. International Journal of Meteorology 32:193–200

Barry, J. (1980) Ball lightning and bead lightning: Extreme forms of atmospheric electricity. Springer Science & Business Media

Berger, K. (1973) Kugelblitz und Blitzforschung. Naturwissenschaften 60:485–492

Boerner, H. (2016) Analysis of conditions favorable for ball lightning creation. 33rd International Conference on Lightning Protection (ICLP) IEEE, (pp. 1–6).

Boissonnat, G. et al (2016) Measurement of ion and electron drift velocity and electronic attachment in air for ionization chambers. arXix preprint:arXiv:1609.03740

Brand, W. (1923) Der Kugelblitz. Henry Grand Hamburg

Brand, W. (1971) Ball Lightning. NASA TT-F13,228

Brand, W. and Wittmann, A. (2010) Der Kugelblitz. Verlag Norbert Kessel

Bychkov, A. V., Bychkov, V. L., and Abrahamson, J. (2002) On the energy characteristics of ball lightning. Philosophical Transactions of the Royal Society of London A 360:97–106

Cameron, R. P. (2018) Monochromatic knots and other unusual electromagnetic disturbances: light localized in 3D. Journal of Physics Communications 2:015024

Campbell, S. (1993) Comment on ball-lightning and greenhouse-effect papers. Journal of Meteorology 18: 259

Campbell, S., (2008) The Case Against Ball Lightning, https://www.skeptic.com/eskeptic/09-12-23/#feature

Cen, J., Yuan, P., and Xue, S. (2014) Observation of the optical and spectral characteristics of ball lightning. Physical Review Letters 112:035001

Chubykalo, A. E. and Espinoza, A. (2002) Unusual formations of the free electromagnetic field in vacuum. Journal of Physics A: Mathematical and General 35:8043

Chubykalo, A. E., Espinoza, A., and Kosyakov, B. P. (2010) Self-dual electromagnetic fields. American Journal of Physics 78:858–861

Cooray, G. and Cooray, V. (2008) Could some ball lightning observations be optical hallucinations caused by epileptic seizures, Open Atmospheric Science Journal, pp 101–105

Cooray, V. (2015) An Introduction to Lightning, Springer, Heidelberg

Cummins, K. L. and Murphy, M. J. (2009) An overview of lightning locating systems: History, techniques, and data uses, with an in-depth look at the US NLDN. Transactions on Electromagnetic Compatibility 51:499–518

Dawson, G. A. and Jones, R. C. (1969) Ball lightning as a radiation bubble. Pure and Applied Geophysics 75:247–262

Decharme, C. (1884) Orage et coup de foudre en fragments globulaire. La Lumière électrique XI:551

Dwyer, J. R. et al (2005) X-ray bursts associated with leader steps in cloud-to- ground lightning. Geophysical Research Letters 32

Dwyer, J. R. and Uman, M. A. (2014) The physics of lightning. Physics Reports 534:147–241

Endean, G. (1997) Development of the radiation bubble model of ball lightning. Journal of Meteorology 22:98–105

Fantz, U. and Friedl, R. and Briefi, S. (2015) Correlation of size, velocity, and autonomous phase of a plasmoid in atmosphere with the dissipated energy. Journal of Applied Physics 117:173301

Grigor'ev, A. and Grigor'eva, I. D. and Shiryaeva, S. (1992) Ball lightning penetration into closed rooms: 43 eyewitness accounts. Journal of Scientific Exploration 6:261–279

Haldoupis, C. et al (2013) The VLF fingerprint of elves: Step-like and long-recovery early VLF perturbations caused by powerful +- CG lightning EM pulses. Journal of Geophysical Research: Space Physics 118:5392–5402

Handel, P. H. and Leitner, Jean-F. (1994) Development of the maser-caviton ball lightning theory. Journal of Geophysical Research: Atmospheres 99:10689–10691

Hare, B. M. et al (2018) LOFAR lightning imaging: Mapping lightning with nanosecond precision. Journal of Geophysical Research: Atmospheres 123:2861–2876

Harrison, R. G. et al (2010) Self-charging of the Eyjafjallajökull volcanic ash plume. Environmental Research Letters 5:024004

Heidler, F. and Diendorfer, G. and Zischank, W., (2004) Examples of severe destruction of trees caused by lightning. 27th International Conference on Lightning Protection, Avignon, France, 8a

Holle, R. L. et al (1997) An isolated winter cloud-to-ground lightning flash causing damage and injury in Connecticut. Bulletin of the American Meteorological Society 78:437–442

Humphreys, W. J. (1936) Ball Lightning. Proceedings of the American Philosophical Society 76:613–626

Idone, V. P., Orville, R. E. and Henderson, R. W. (1984) Ground truth: A positive cloud-to-ground lightning flash. Journal of climate and applied meteorology 23:1148–1151

Jennison, R. C. (1969) Ball lightning. Nature 224.5222: 895.

Jennison, R. C. (1990) Relativistic phase-locked cavity model of ball lightning. Physical Interpretations of Relativity Theory: Proceedings 2:359

Kammer, T. et al (2005) Transcranial magnetic stimulation in the visual system. II. Characterization of induced phosphenes and scotomas. Experimental Brain Research 160:129–140

Keul, A. and Stummer, O. (2002) Comparative analysis of 405 Central European ball lightning cases. Journal of Meteorology 27:385–393

Keul, A. G., (2008) European ball lightning statistics. Proceedings of the 10th International Symposium on Ball Lightning (ISBL08) and 3rd International Symposium on Unconventional Plasmas (ISUP08), Kaliningrad, Russia

Keul, A. G. and Sauseng, P. and Diendorfer, G. (2008) Ball lightning - An electromagnetic hallucination? International Journal of Meteorology 33:89–95

Keul, A. G. and Diendorfer G. (2018) Assessment of ball lightning cases by correlated LLS data. 34th International Conference on Lightning Protection (ICLP)

Lowke, J. J. (1996) A theory of ball lightning as an electric discharge. Journal of Physics D: Applied Physics 29:1237

Lowke, J. J. et al (2012) Birth of ball lightning. Journal of Geophysical Research: Atmospheres 117

Lyons, W. A. and Uliasz, M. and Nelson, T. E. (1998) Large peak current cloud-to-ground lightning flashes during the summer months in the contiguous United States. Monthly Weather Review 126:2217–2233

MacGorman, D. R. and Rust, W. D. (1998) The Electrical Nature of Storms, Oxford University Press

McNally Jr, J. R. (1966) Preliminary report on ball lightning, Oak Ridge National Lab.

O'Reilly, B. (1962) Green Mountains and Cullenbenbong, Qld. Book Depot

Ohtsuki, Y.H. and Ofuruton, H. (1991) Plasma fireballs formed by microwave interference in air. Nature 350:139

Panayiotopoulos, C. P. (1999) Elementary visual hallucinations, blindness, and headache in idiopathic occipital epilepsy: differentiation from migraine. Journal of Neurology, Neurosurgery & Psychiatry 66:536–540

Pedeboy, S. et al (2017) Characteristics and distribution of intense cloud-to-ground flashes in Western Europe. International Colloquium on Lightning and Power Systems Ljubljana

Peer, J. and Cooray, V. and Cooray G. and Kendl, A. (2010) Erratum and addendum to "Transcranial stimulability of phosphenes by long lightning electromagnetic pulses" [Phys. Lett. A 374 (2010) 2932]. Physics Letters A 374:4797–4799

Pierce, E. T. et al (1960), An Experimental Investigation of Negative Point-plane Corona and Its Relation to Ball Lightning, AVCO corporation RAD TR-60-29

Rakov, V. A. and Uman, M. A. (2003) Lightning: physics and effects, Cambridge University Press

Rakov, V. A. (2003) A review of positive and bipolar lightning discharges. Bulletin of the American Meteorological Society 84:767–776

Ranada, A. F. and Soler, M. and Trueba, J. L. (2000) Ball lightning as a force-free magnetic knot. Physical Review E 62:7181

Rayle, W. D (1966) Ball Lightning Characteristics. NASA TN D-3118

Saba, M. M. F., et al. (2017) Lightning attachment process to common buildings. Geophysical Research Letters 44: 4368–4375.

Said, R. K. and Cohen, M. B. and Inan, U. S. (2013) Highly intense lightning over the oceans: Estimated peak currents from global GLD360 observations. Journal of Geophysical Research: Atmospheres 118:6905–6915

Sagan, P. (2004) Ball Lightning: Paradox of Physics. iUniverse, Inc.

Singer, S. (1971) The Nature of Ball Lightning. Plenum Press

Singer, S. (2002) Ball lightning - the scientific effort. Philosophical Transactions: Mathematical, Physical and Engineering Sciences 360:5–9

Smirnov, B. M. (1987) The properties and the nature of ball lightning. Physics Reports 152:177–226

Stepanov, S. I. (1990) On the Energy of Ball Lightning. Sov. Phys. Tech. Phys. 35:267

Stephan, K. D. (2012) Implications of the visual appearance of ball lightning for luminosity mechanisms. Journal of Atmospheric and Solar-Terrestrial Physics 89:120–131

Stephan, K. D. (2016) Extension of Relativistic-Microwave Theory of Ball Lightning Including Long-Term Losses and Stability. arXiv preprint arXix:1608.00450

Stenhoff, M. (1999) Ball Lightning. An Unsolved Problem in Atmospheric Physics. Kluver Academic/Plenum Publishers

Thomas, R. J. et al (2007) Electrical activity during the 2006 Mount St. Augustine volcanic eruptions. Science 315:1097–1097

Thomas, R. J. et al (2010) Polarity and energetics of inner core lightning in three intense North Atlantic hurricanes. Journal of Geophysical Research: Space Physics 115

Thornton, J. A. et al (2017) Lightning enhancement over major oceanic shipping lanes. Geophysical Research Letters 44:9102–9111

Tompkins, D.R. and Rodney, P.F. (1980) Possible photographic observations of ball lightning. Il Nuovo Cimento C 3:200–206

Torchigin, V. P. and Torchigin, A. V. (2004) Behavior of self-confined spherical layer of light radiation in the air atmosphere. Physics Letters A 328:189–195

Turner, D. J. (1998) Ball lightning and other meteorological phenomena. Physics reports 293:2–60

Versteegh, A. et al (2008) Long-living plasmoids from an atmospheric water discharge. Plasma sources science and technology 17:024014

Wang, D., Kuroda, S. and Takagi, N. (2016) Lightning attachment process parameters measured by using LAPOS. Conference: 2016 33rd International Conference on Lightning Protection (ICLP)

Williams, E. et al (2012) Resolution of the sprite polarity paradox: The role of halos. Journal of Geophysical Research: Space Physics 47 LAPOS. 33rd International Conference on Lightning Protection (ICLP), Estoril

Wu, H.-C. (2016) Relativistic-microwave theory of ball lightning. Scientific reports 6:28263

Zheng, Xue-Heng (1990) Quantitative analysis for ball lightning. Physics Letters A 148:463–469

Printed in the United States
By Bookmasters